Promotors: **Prof. dr. ir. Mieke Uyttendaele**
Em. Prof. dr. ir. Johan Debevere
Department of Food Safety and Food Quality
Faculty of Bioscience Engineering, Ghent University

Dean: **Prof. dr. ir. Guido Van Huylenbroeck**

Rector: **Prof. dr. Paul Van Cauwenberge**

ir. Vicky Jasson

Resuscitation and enrichment of food-borne pathogens as sample preparation step to serve fast detection methods

Thesis submitted in fulfillment of the requirements
For the degree of Doctor (PhD) in Applied Biological Sciences

Resuscitatie en ophoping van voedselgebonden pathogenen als monstervoorbereidingsstap bij het gebruik van snelle detectiemethoden.

Illustration cover: A cartoon representing the resuscusitation and enrichment steps of food-borne pathogens with respect to fast detection methods prepared by Johan Desseyn and Vicky Jasson.

To refer to this thesis:

Jasson, V. 2010. Resuscitation and enrichment of food-borne pathogens as sample preparation step to serve fast detection methods. PhD dissertation, Faculty of Bioscience Engineering, Ghent University, Ghent.

ISBN: 9789059893566

Woord vooraf

Voedselveiligheid het boeide mij meteen! Bij de voorstelling van het doctoraatsonderwerp over resuscitatie en ophoping van voedselgebonden pathogenen, kaderend binnen Pathogen Combat, heb ik geen moment getwijfeld. Voor dit onderwerp wilde ik echt gaan! Het was echter niet evident om in een project met een aantal vooropgestelde mijlpalen en deadlines persoonlijk, maar vooral publicatiewaardig werk neer te zetten. In mijn middelbare-school-carrière zeiden verschillende leerkrachten: “Vicky je maakt het jezelf nooit gemakkelijk!” maar ik denk nu dat juist deze eigenschap heeft bijgedragen tot het persoonlijke en het publicatiewaardige van dit doctoraat. Het zoeken naar lage aantallen subletaal beschadigde cellen is “zoeken op de limiet”, nog maar te zwijgen van verborgen *Campylobacter* spp. Dit boekje heeft bloed, zweet en tranen gekost, maar zit ook vol optimisme, overwonnen uitdagingen en vooral veel werkplezier en dit mag je letterlijk nemen. Aangezien wetenschappelijk onderzoek is iets dat je niet alleen doet, wil ik graag een aantal mensen bedanken.

Mijn oprechte dank gaat uit naar mijn promotoren Em. Prof. Johan Debevere en Prof. Mieke Uyttendaele die mij samen de kans hebben gegeven om te starten aan dit doctoraat. Prof. Debevere, ik dank u voor het vertrouwen dat u stelde in mij als onderzoeker en als medewerker van uw toenmalig labo. Ik heb mijn best gedaan om te passen binnen de LFMFP-groep en om mee te fungeren als uithangbord van het labo. Prof. Uyttendaele, Mieke, u wil ik in het bijzonder bedanken voor de steun, het vertrouwen, de samenwerking en de kritische zin gedurende het volledige “doctoraatsparcours”. Ik ben vol bewondering voor uw kennis, enthousiasme en gedrevenheid die mij persoonlijk gemotiveerd en geïnspireerd hebben om mijn ziel in dit werk te leggen. Jij hebt mij als wetenschapper doen groeien en ik heb mij altijd héél erg goed gevoeld onder uw vleugels! Merci!

Prof. Frank Devlieghere en dr. Liesbeth Jacxsens wil ik bedanken voor hun kritische blik op een aantal gestelde problemen die zich voordelen tijdens mijn onderzoek. Dankzij onze verschillende invalshoeken vonden we de oplossing. Dr. Rajkovic, of beter “Professor Andreja”, bedankt dat je er steeds was voor mij zowel persoonlijk als wetenschappelijk, zowel tijdens de werkweek maar ook in het weekend! Gelukkig (voor mij) deelden wij de ganse tijd eenzelfde bureau en had ik steeds snel antwoord op al mijn vragen. Ik heb echt héél erg veel van jou geleerd!

Oprecht wil ik ook alle collega’s bedanken. Zij hebben steeds begrip getoond voor de grote experimenten die ik uitgevoerde en het dus bijhorende plaatsgebrek. Katrien Vanwildemeersch en An De Coen wil ik in het bijzonder bedanken voor de ondersteuning bij praktische vragen en hun hulp bij verschillende experimenten. Het was echt aangenaam werken, zelfs bij epje 299...

Nada Smigic and An Vermeulen, I would like to thank you both as colleagues but most of all as human beings! An, waarschijnlijk besef jij het niet, maar dankzij jou heb ik beter leren relativeren! Onze gesprekken op de trein, die vergeet ik nooit. Nada, I would like to thank you for so many thinks, but most of all I would like to say: “because of you, I could be me, myself and I during the working hours!” This was the biggest support of all! Hvala!

Ook Prof. Lieven De Zutter wil ik graag bedanken voor het ter beschikking stellen van zijn labo voor de vergelijkende studie van de IMS methoden. Lieven bedankt voor uw kritisch kijk op mijn experimenten en de steeds zeer snelle “replies”op alle gestelde vragen. Karen en Annelies ik vond het zeer aangenaam om met jullie samen te werken.

Met de thesisstudenten liep niet alles van een leien dakje tijdens mijn “doctoraatsparcours”, maar twee van hen heb ik echt veel gehad. Paco, I would like to thank you for the long working days regarding the Bolton-Preston experiments of *Campylobacter* which you always took with a smile and with politeness. Also many thank to Anna how performed a nice start of the study regarding alternative method in microbiology.

Deze bundeling van onderzoeksresultaten is nauwgezet gelezen door de leden van de examencommissie. De geleverde commentaren en suggesties hebben dit proefschrift aanzienlijk verbeterd. Heel erg bedankt Dr. ir. Koen De Reu, Dr. ir. Paul In’t Veld, Dr. Geert Huys en Prof. Lieven De Zutter.

Graag wil ik ook mijn familie bedanken. Mijn grootouders en ouders voor de vorming die jullie mij gaven die mij maakte tot wie ik ben en waardoor het mogelijk was om dit boekje neer te leggen. Mijn zus Evi en mijn mama Isa wil ik speciaal bedanken voor hun begrip en hun kalmerende woorden tijdens de stressmomenten. Ook veel dank aan Margriet en Mon voor hun luisterend oor!

Mijn vrienden, Annelies, Sarah, Astrid en Hans, Brigitte, Katrien en Josefien. Bedankt voor de steun en de aanmoedigingen, maar vooral voor de vele ontspannende feestjes en de soms nodige “babbels”. Misschien komt er nu een tijd dat ik echt ontspannen zal zijn.

Als laatste, maar daarom zeker niet de minste dank gaat uit naar “mijne Nikki”. Ik weet dat ik niet de gemakkelijkste persoon ben. Door mijn drive voor mijn werk werd dit wel eens uitvergroot de laatste jaren, maar de manier waarop je hiermee omging was fantastisch! Ik besef dit maar al te goed! Jij krijgt mijn scherpe kantjes eraf en daar ben ik heel gelukkig mee! Ik wil je vooral bedanken voor het wachten. Het wachten om samen leuke dingen te doen, het wachten op het “leven”. Wij hebben vele dingen gelaten en andere nog niet kunnen waarmaken, maar ik hoop dat er nu verandering in zal komen. Merci om er te zijn voor mij! Jij bent mijn voet in het echte leven! Ik hoop dat de stress nu een beetje kan luwen zodat ik een iets meer relaxt en nog spontaner naast je kan staan.

Op naar het volgende!

Vicky Jasson

Gent, februari 2010

Table of contents

Table of contents

List of abbreviations

ADP	Adenosine diphosphate
AFNOR	Association Français de Normalisation
AK	Adenylate kinase
ALOA	Agar *Listeria* Ottaviani & Agosti
Amp	Ampicillin
ANOVA	Analysis of variance
AOAC	Association of Official Analytical Chemist
ATP	Adenosine triphosphate
A_w	Water activity
BCCM	Belgian co-ordinated collections of microorganisms
BPW	Buffered peptone water
BPW+SM+BG	Buffered peptone water with skimmed milk powder and brilliant green
BPW+V	BPW with vancomycin
BSL	Bio-Safety level
bTSBYE	Buffered tryptone soy broth with yeast extract
CBA	Colombia agar base
CCDA	Charcoal cefoperazone deoxycholate agar
CFU	Colony forming unit
Chl	Chloramphenicol
Cip	Ciprofloxacin
CLSI	Clinical and Laboratory Standards Institute
Col	Colistin
Ct	Cycle threshold
CT	Cefixime tellurite
CT-SMAC	Sorbitol MacConkey agar with cefixime tellurite
D	Detection
DNA	Deoxyribonucleic acid
E	Enrichment
EE	Extra enrichment
ECred	EC with reduced Bile salts

ELISA	Enzyme linked fluorescent assay
ESBL	Extended-spectrum-beta-lactamase
FDA	Fluorescein diacetate
FISH	Fluorescence *in situ* hybridization
Flor	Florphenicol
Fot	Cefotaxime
FSMS	Food safety management system
GBS	Guillain-Barré syndrome
Gm	Gentamicin
GMP	Good manufacturing practice
GT	Generation time
HACCP	Hazard analyses and critical control points
HC	Hemorrhagic colitis
HHP	High hydrostatic pressure
HUS	Haemolytic uraemic syndrome
IAC	Internal amplification control
ILP	Intensive light pulse
IMS	Immunomagnetic separation and concentration
ISO	International Standards Organization
Kan	Kanamycin
KVE	Kolonie vormende eenheid
LFD	Lateral flow devices
LFMFP	Laboratory of Food Microbiology and Food Preservation
LMG	Laboratory of Microbiology Ghent
LRB	*Listeria* repair broth
LSB	*Listeria* special broth
LX	*Listeria* Xpress broth
MA	Microbial analysis
MAS	Microbial assessment scheme
mCCDA	Modified charcoal cefoperazone deoxycholate agar
MIC	Minimum inhibitory concentration
MKTTn	Mueller-Kauffmann tetrathionate novobiocin broth
MO	Microorganism
MPCR	Multiplex PCR

mRNA	Messenger RNA
mTSB+N	modified Tryptone soy broth with novobiocin
MUG	Methylumbrelliferyl-D-glucuronidase
NA	Not applicable
Nal	Nalidixic acid
NB	Nutrient broth
ND	Not determined
NS	Non-selective medium
OD	Optical density
OEB	One enrichment broth
OMA	Official method of analysis
PCA	Plate count agar
PCR	Polymerase chain reaction
PE	Pre-enrichment
PEMC	Piezoelectric excited millimeter-sized cantilever
PI	Propidium iodide
PPS	Peptone physiological saline
PTM	Performance tested method
Rep	Repetition
RM	Reference material
RNA	Ribonucleic Acid
rRNA	Ribosomal RNA
RTE	Ready-to-eat
RVS	Rappaport-Vassiliadis soya peptone broth
S	Selective medium
STEC	Shiga-like toxin producing *E. coli*
Str	Streptomycin
SUL	Sulphametoxazole
SX	*Salmonella* Xpress broth
Taz	Ceftazidime
TBX	Tryptone-bile-glucuronic
Tet	Tetracyclin
Trmp	Trimethoprim
TSA	Tryptone soy agar

TSAYE	Tryptone soy agar with yeast extract
TSAYE-NaCl	Salted tryptone soy agar with yeast extract
TSB	Tryptone soy broth
TSB+	Tryptone soy broth with novobiocin and vancomycin
TSB++	TSB with novobiocin, vancomycin, rifampycin, K-telluriet and Bile salts *N°3*
TSI	Triple sugar iron
TTD control	Time to detection of healthy cells
TTD stress	Time to detection of stressed cells
TTP	Thrombotic thrombocytopenic purpura
VIT	Vermicon Identification Technology
VRBL	Violet red bile lactose agar
VT	Verocytotoxin
VTEC	Verocytotoxin producing *E. coli*
XLD	Xylose lysine deoxycholate
YE	Yeast extract
YOPI	Young, old, pregnant and immunocompromised people

Introduction and objectives

Introduction and objectives

More than 200 zoonotic diseases are spread through food and millions of people fall ill every year and die due to the food-borne illness. While most food-borne illnesses are sporadic and often not reported, food-borne outbreaks may take on massive proportions. In industrialized countries, the percentage of the population suffering from food-borne illness each year has been reported to be up to 30%. While less well documented, developing countries bear the brunt of the problem due to the presence of a wide range of food-borne illnesses.

Chemicals, heavy metals, parasites, fungi, viruses and bacteria can cause food-borne illness. Bacteria related food poisoning is the most common, but fewer than 20 of the many thousands of different bacteria are the culprits. More than 90 percent of the cases of food poisoning each year are caused by *Staphylococcus aureus, Salmonella* spp., *Clostridium perfringens, Campylobacter* spp., *Listeria monocytogenes, Vibrio parahaemolyticus, Bacillus cereus,* and Entero-pathogenic *Escherichia coli.* These bacteria are commonly found on many raw foods (www.WHO.int, 2010).

A difference is made between bacteria causing food-borne infection such as *L. monocytogenes* and bacteria causing food-borne intoxication such as *S. aureus.* Food-borne infection refers to the presence of bacteria which infect the body after consumption of the food, while food-borne intoxication refers to the ingestion of toxins contained within the food. Although it should be noticed that also infection-mediated toxicity exists.

S. aureus produces toxins that cause intense vomiting in humans and are the major cause of food-borne outbreaks associated with bacterial toxins. *C. perfringens* bacteria multiply especially in food prepared from red meat and poultry and its toxins cause abdominal cramps and diarrhea. *Bacillus cereus* may produce emetic and diarrheagenic toxins. Depending on the type of toxin, *B. cereus* may cause severe nausea, vomiting and watery diarrhoea. Most bacterial toxins can be destroyed by normal heat treatment (exceptions include certain staphylococcal enterotoxins and the emetic toxin of *B. cereus*). *Clostridium botulinum* toxin is the cause of a rare but potentially deadly intoxication and occurs when the anaerobic bacterium grows in foods and produces botulinum toxin, a powerful paralytic toxin (EFSA, 2009).

Some food-borne infections are well recognized, but are also considered emerging because they have recently become more common. For example, outbreaks of *Salmonella* spp. have been reported for decades all over the world, but within the past 25 years the disease has increased in incidence on many continents (e.g. the Western hemisphere and Europe). While cholera, a disease caused by *Vibrio cholerae* has devastated much of Asia and Africa for years (tropical regions), its introduction for the first time in almost a century on the South American continent in 1991 makes it another example of an infectious disease that is both well-recognized and emerging. While cholera is often water-borne, many foods also transmit infection also causing enormous economic losses (rice, vegetables, millet gruel and various types of seafood). In Latin America, ice and raw or under-processed seafood are important epidemiological pathways for cholera transmission. *Vibrio parahaemolyticus*, another pathogenic bacterium from the genus vibrio, is also found on seafood, and requires the salt environment of sea water for growth. *V. parahaemolyticus* is very sensitive to cold and heat. Food poisoning caused by this bacterium is a result of insufficient cooking and/or contamination of the cooked product by a raw product, followed by improper storage temperature. It is a major problem in Japan where many seafoods are consumed raw. *Vibrio vulnificus* is another member of the vibrio genus that is found in the marine environment. *V. vulnificus* is truly an emerging pathogen, but it can be controlled with proper cooking and refrigeration.

Even though *Yersinia enterocolitica* is not a frequent cause of human infection in the temperate climate regions (not recognised as emerging), it can be cause of human infection, mostly in the cool climate regions, and it is often involved in illness with very severe symptoms. Yersiniosis, infection caused by this microorganism, occurs most commonly in the form of gastroenteritis. Children are most severely affected. Symptoms of pseudoappendicitis resulted in many unnecessary appendectomies. Death is rare and recovery is generally complete in 1-2 days. *Y. enterocolitica* is commonly present in foods, but with the exception of pork, most isolates do not cause disease (Fredriksson-Ahomaa, 2006).

Other food-borne pathogens are considered emerging because they are new microorganisms or because the role of food in their transmission has been recognized only recently. Infection with *Escherichia coli* O157:H7 was first described in 1982. Subsequently, it has emerged rapidly as a major cause of bloody diarrhoea and acute renal failure. Outbreaks of infection, generally associated with beef, have been reported in Australia, Canada, Japan, United States, in various European countries, and in southern Africa. *L. monocytogenes* is considered emerging because the role of food in its transmission has only been recognized since 1981. Outbreaks of *L. monocytogenes* have been reported from many countries, including Australia, Switzerland, France and the United States.

Campylobacter spp. are widespread and in some countries especially the European ones, the reported number of cases surpasses the incidence of *Salmonella* infection.

L. monocytogenes, *Campylobacter* spp., *Salmonella* spp. and *E. coli* O157:H7 are four major food-borne pathogens, also referred to as the "big 4". These pathogens are important in governmental food inspection to enforce legal regulations, in international trade to determine compliance with microbiological standards, in commercial relationships between trade partners to control on agreed microbiological specifications, in the food industry to verify the well-functioning of the food safety management system (FSMS), in academic laboratories conducting research studies and in reference laboratories involving in monitoring and baseline surveys.

There is a continuous desire for rapid methods especially in the framework of food-borne outbreaks and compliance testing. A short time to detection enables immediate decisions on acceptance or rejections and corrective measures e.g. positive batch release. Besides, during the last years, people prefer fresh and healthy foods while preservatives are becoming less popular. Without preservatives the shelf-life of the food product becomes shorter, and with conventional detection methods (e.g. the ISO methods) the time required to ensure the absence of pathogens would be excessive. It is clear that a short shelf-life (e.g. raw fresh or minimal processed foods) is proportional with a high priority to set a short time to detection. However, the rapid method should be reliable i.e. specific (no false positive results) and sensitive (no false negative results). The availability of rapid reliable methods will contribute significantly to guarantees on food safety.

The **first objective** of this PhD study was to give a detailed overview of the different alternative **rapid detection methods** for *L. monocytogenes*, *E. coli* O157, *Campylobacter* spp. and *Salmonella* spp., which is presented in **Chapter 1**. An overview of the alternative detection methods of these pathogens in food, in particularly the validated methods, is given with reference to the conventional ISO methods.

Due to the low infectious dose of *L. monocytogenes*, *E. coli* O157, *Salmonella* spp. and *Campylobacter* spp. (10-1000 cells), the number and frequency of the reported outbreaks and the severity of the illnesses of these major 4 pathogens, the detection methods in foods need a low detection limit (<10 CFU/25g). Food, due to processing (freezing, heat-treatment, etc.) and preservation techniques (salting, acidification, packaging, etc.) presents non-optimal conditions for microorganisms. As a consequence the pathogens are stressed and/or sub-lethally injured. If advanced rapid detection methods overlook the need for appropriate resuscitation and enrichment then this may result in a significant risk of false negative results. The validation of rapid alternative methods should therefore consider the detection of sub-lethally injured cells as well. For this purpose it is recommended to set up protocols that can provoke injury enabling simulations of the status of the natural flora of naturally contaminated samples. Therefore the **second objective** was to determine **standardized procedures for creating sub-lethally injured cells**, which is shown in **Chapter 2**. This chapter provides a comprehensive and structured examination on the effect of various food-related stress conditions on the ability to recover these food-borne pathogens.

The **third objective** was to evaluate the **kinetics of resuscitation and enrichment** and to determine **optimum conditions for resuscitation and enrichment** of also sub-lethally injured pathogens to serve as sample preparation for detection by alternative rapid methods. These results, obtained for respectively *L. monocytogenes*, *E. coli* O157, *C. jejuni* and *S. enterica*, are presented in **Chapters 3, 4, 5 and 6**. Appropriate resuscitation and enrichment conditions need to favour repair of sub-lethally injured target cells and to enable multiplication of the target pathogen in order to obtain the numbers needed for detection. Apart from appropriate culture to increase the levels of pathogens, also concentration of pathogens in a small volume free of interfering components is often a prerequisite to facilitate the technical performance of the rapid methods.

The research outline of this PhD study is presented in **Fig. 0.1**. These studied objectives are aimed to give insight in the bottlenecks, especially associated with sample preparation, existing in the development of alternative rapid detection methods to be applied in food diagnostics. Furthermore this PhD study wants to offer some selection criteria for alternative rapid detection methods which are presented in Chapter 7, the discussion part of this PhD dissertation.

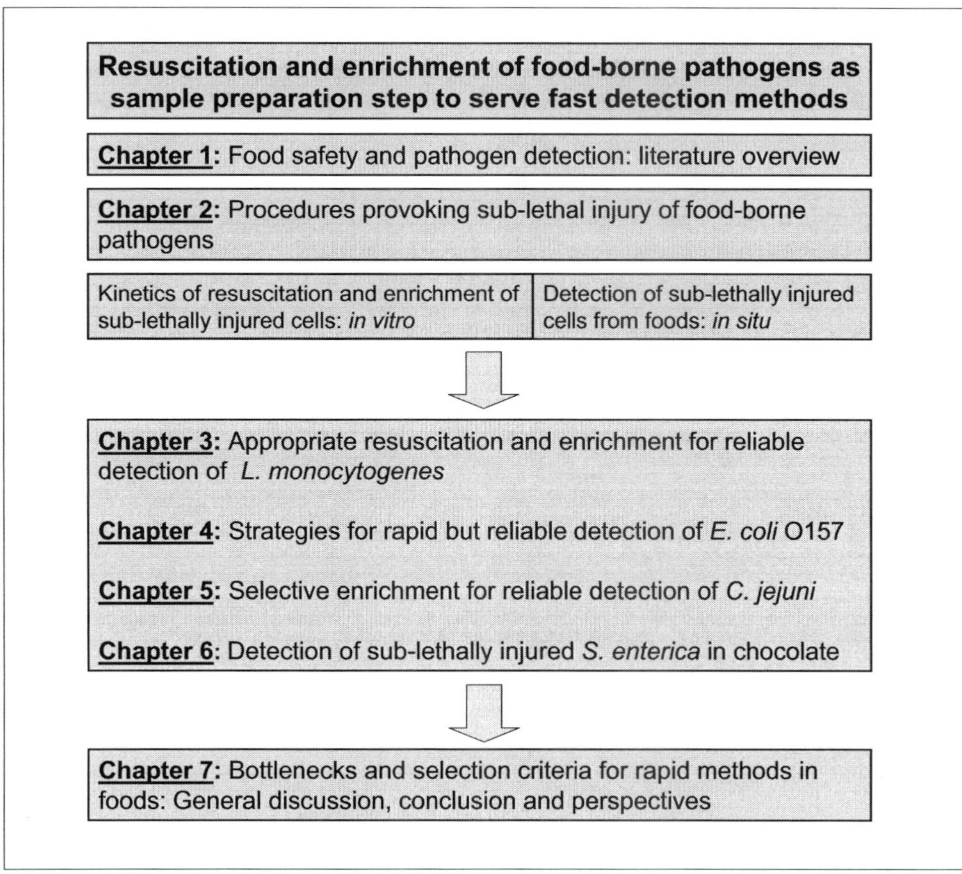

Fig. 0.1. Research outline of this PhD study.

Summary - Samenvatting

Summary

The literature review presented in **Chapter 1** describes, on the one hand, the current state-of-the-art regarding rapid detection methods for *Listeria monocytogenes*, *Escherichia coli* O157, *Campylobacter* spp. and *Salmonella* spp., and on the other hand, it addresses the different bottlenecks of these rapid detection methods. In the first place specific characteristics of the four above mentioned pathogens, their significance and the most related food product are described. Secondly a description of the conventional ISO methods is given followed by an overview of the rapid alternative detection methods, in particularly the validated detection methods. Specific bottlenecks of these rapid detection methods are summarised in the third part the literature study, such as the presence of sub-lethally injured cells. The last part addresses the importance of sample preparation for reliable detection of food-borne pathogens.

As this PhD research focused in particularly on the detection of sub-lethally injured cells in food, **Chapter 2** describes methods to create sub-lethally injured cells in a standardized manner. The aim of the study presented in this chapter was to create a model for controlled injury and methods for the determination of injury and repair of three pathogens: *L. monocytogenes*, *E. coli* O157 and *Campylobacter jejuni*. Therefore the pathogens were exposed to several stress conditions (heat, acid, freezing, cold, oxidative and food stress) mimicking conditions often used as preservation techniques. A methodology to determine sub-lethal injury was established and threshold values indicating the significance of the percentage of sub-lethally injured cells due to treatment were obtained. For *L. monocytogenes* the freeze stress, the oxidative and the food stress (mimicking conditions of soft cheese) were able to create a significant amount of sub-lethally injured cells, while for *E. coli* O157, the freeze stress, the oxidative stress and the food stress (mimicking a fermented sausage) were capable of creating significant amounts of sub-lethally injured cells. For *C. jejuni* a specific “food” stress (salted minced chicken meat) as well as the oxidative stress were shown to be standardized procedures provoking sub-lethal injury. The standardized procedures, resulting in sub-lethally injured cells could be further used for performance testing of resuscitation and enrichment broths.

In **Chapter 3**, a study was performed on various non-selective and newly introduced selective enrichment broths regarding their resuscitation and enrichment capacities for *L. monocytogenes*. First, the growth parameters lag phase (λ) and maximum specific growth rate (μ_{max}) of both healthy and sub-lethally injured cells were determined. Next, a selection of enrichment broths was compared for their capacity to support resuscitation and enrichment of low numbers of *L. monocytogenes* in artificially and naturally contaminated food samples within 24 h. Detection was performed either by phage protein-based capture (*Listeria* Capture kit, Profos, Regensburg, Germany) combined with plating on chromogenic medium or by fluorescence *in situ* hybridization (FISH) using the VIT-*Listeria* kit (Vermicon, Munich, Germany). Kinetics of resuscitation and growth of *L. monocytogenes* in various enrichment broths showed that for detection of low numbers of sub-lethally injured *L. monocytogenes* cells at least an overnight enrichment was needed. The use of a non selective enrichment broth, although very promising *in vitro*, was shown to be not appropriate for *L. monocytogenes* detection in foods. A moderate selective enrichment broth was needed to enable proliferation of the target cells in the presence of indigenous bacterial flora. However, combination of an appropriate enrichment condition with advanced detection techniques may enable a reliable 24 h detection of *L. monocytogenes*.

In **Chapter 4**, a comparative study of λ and μ_{max} of healthy, stressed and sub-lethally injured *E. coli* O157 in ten enrichment broths was performed. The evaluation of enrichment protocols was validated by different end-point detection methods. Two real-time PCR methods, iQ-Check™ *E. coli* O157:H7 PCR kit (Bio-Rad, Nazareth Eke, Belgium) and TaqMan® *E. coli* O157:H7 PCR kit (Applied Biosystems, Lennik, Belgium), and two combined capture/plate methods *E. coli* O157 Anti-O157 Dynabeads (Dynal, Oslo, Norway) and *E. coli* O157 phage protein-based capture Kit (Profos) were used as detection methods. TSB (3) provided the fastest growth (μ_{max} = 1.00 $\pm$ 0.06 h^{-1}) but failed to recover oxidative stressed *E. coli* O157. TSB (1), TSBYE, TSB supplemented with 8 mg/l novobiocin + 16 mg/l vancomycin (TSB+), BPW and BPW supplemented with 8 mg/l vancomycin (BPW+V) enabled resuscitation of *E. coli* O157 independent from pre-cultural conditions. mTSB+N, EC, ECred, TSB (3) and TSB supplemented with 8 mg/l novobiocin + 16 mg/l vancomycin

+ 2 mg/l rifampycin + 1 mg/l K-Telluriet + 1.5 g/l Bile salts *N°3* (TSB++) all failed recovery of *E. coli* O157 for at least one stress. The use of TSB (1), TSB+, BPW and BPW+V was compared to mTSB+N, the ISO 16654:2001 (International Standards Organization, 2001a) recommended enrichment broth, for reliable detection from different foods (sprouted seeds, fermented sausage, raw milk, raw ground beef). When low numbers (approximately 1 CFU/g) of healthy cells were inoculated, BPW/BPW+V, TSB, TSB+ and mTSB+N enabled growth until detectable numbers within 6 hours enrichment at 41.5°C. Results showed that mTSB+N failed to recover freeze and "food" sub-lethally injured *E. coli* O157 until detectable numbers in contrast to BPW/BPW+V. This study highlights that using mTSB+N as indicated by ISO 16654:2001 for recovery of *E. coli* O157 from foods may yield false negative results.

In **Chapter 5** a comparative study of Bolton and Preston as enrichment broths for reliable detection of *C. jejuni* was performed. The experimental setup included both healthy and freeze stressed *C. jejuni* cells. Tested as pure cultures, Bolton broth enabled faster resuscitation and growth of *C. jejuni* compared to Preston broth. When *C. jejuni* was co-incubated with extended-spectrum-beta-lactamase (ESBL) producing *Escherichia coli* isolated from Belgian poultry meat preparations, the latter dominated in the Bolton enrichment broth and overcrowded the mCCDA isolation plates. This resulted in the inability to recover *C. jejuni* by ISO 10272:2006 (International Standards Organization, 2006b). Preston broth required a longer incubation time to detect *C. jejuni* compared to Bolton broth. However, in Preston broth, the ESBL *E. coli* isolates failed to grow. The use of the same antibiotic (sodium cefoperazone) in Bolton broth and in mCCDA plates may explain the problems encountered for detection of *C. jejuni*, as ESBL *E. coli* are resistant to this antibiotic. The ESBL *E. coli* caused overgrowth and masked the few *C. jejuni* colonies present on the mCCDA plates after enrichment in Bolton broth. The use of *Campylobacter* spp. specific real-time PCR circumvented these problems and enabled rapid detection of the pathogen after 24 h enrichment both in Bolton and in Preston broth, for both healthy and freeze stressed cells either with or without the presence of these ESBL *E. coli* as competitive flora.

In **Chapter 6,** a study was presented that compared different enrichment and DNA extraction protocols to assess their capacity to detect low levels of healthy, but also sub-lethally injured *Salmonella enterica* cells in chocolate as food matrix with real-time PCR. Real-time PCR, iQ-Check™ *Salmonella* II (Bio-Rad), was used as detection assay. VIDAS® Easy SLM (BioMérieux, Marcy l'Etoile, France) and ISO 6579:2005 (International Standards Organization, 2005b) the reference method were used as confirmation methods. Both alternative methods are validated by AFNOR for all foods. In this study the food matrix chocolate was challenged. Two DNA extraction protocols, Standard and Easy extraction both applicable before iQ-Check™ *Salmonella* II, were tested. Easy extraction protocol showed to be more appropriate as DNA extraction method for chocolate as food matrix. Buffered peptone water, the recommended ISO pre-enrichment medium was shown to be incapable to recover low levels of sub-lethally injured *S. enterica* cells from chocolate. It was demonstrated in this study that iQ-Check *Salmonella* II can be used as detection method for naturally contaminated chocolate samples after appropriate sample preparation. All positive samples could be confirmed by another technique as at least 250 CFU/ml of *S. enterica* should be present after enrichment to provide a PCR positive result. Detection of *S. enterica* cells in chocolate by real-time PCR is shown to still be a challenge due to low numbers of sub-lethally injured target cells and inhibition of the PCR reaction.

The knowledge acquired in the above-mentioned chapters has led to the formulation of a general discussion about the bottlenecks of resuscitation and enrichment including selection criteria regarding rapid methods for microbial analysis, which is presented in **Chapter 7**. Also the conclusions obtained from this PhD dissertation and the remaining issues of concern are addressed. The added value and the essence of this PhD were summarized. Finally from the current knowledge and the discussion, perspectives for further research were announced.

Samenvatting

Hoofdstuk 1 geeft een stand van zaken weer betreffende snelle detectiemethoden voor *Listeria monocytogenes*, *Escherichia coli* O157, *Campylobacter* spp. en *Salmonella* spp. In eerste instantie werden specifieke eigenschappen van de verschillende micro-organismen, hun belang voor voedselveiligheid alsook de geassocieerde levensmiddelen beschreven. De conventionele detectiemethoden maar eveneens verschillende alternatieve snelle methoden, meer in het bijzonder de gevalideerde snelle methoden werden onder de loep genomen in een tweede deel van Hoofdstuk 1. Specifieke pijnpunten betreffende de snelle alternatieve methoden werden in kaart gebracht en in een laatste deel werd de link gelegd naar het belang van juiste monstervoorbereiding.

De focus van dit doctoraat lag bij de detectie van subletaal beschadigde cellen uit levensmiddelen. Daarom werden er in **Hoofdstuk 2** protocollen beschreven die op een gestandaardiseerde manier subletaal beschadigde cellen van de pathogenen in kwestie konden creëren. Het doel van dit hoofdstuk was cellen te creëren zoals mogelijk aanwezig in natuurlijk gecontamineerde levensmiddelen. Hiervoor werden de pathogenen blootgesteld aan verschillende niet optimale omgevingssituaties zoals hitte, zuur, invriezen, koude, oxidatieve en een levensmiddel nabootsende situatie. Een methodologie die in staat was om subletale beschadiging te creëren werd vastgelegd. Per pathogeen werd er eveneens een cut-off waarde bepaald die de significantie van de subletale beschadiging kon aantonen. Voor *L. monocytogenes* waren de zure, de oxidatieve en de levensmiddel nabootsende behandeling (zachte kaas) in staat om een significante hoeveelheid subletaal beschadigde cellen te creëren, terwijl voor *E. coli* O157 eveneens de oxidatieve, maar vooral het invriezen en de voor *E. coli* O157 specifiek levensmiddel nabootsende behandeling (gefermenteerde worts) subletaal beschadigde cellen voortbracht. Voor *C. jejuni* was de specifieke levensmiddel nabootsende stress (gezout gehakt kippenvlees) alsook de oxidatieve stress in staat subletaal beschadigde cellen op een gestandaardiseerde manier te creëren. De procedures voorgesteld in Hoofdstuk 2 werden ontwikkeld om de performatie van resuscitatie- en aanrijkingsmedia te kunnen evalueren.

In **Hoofdstuk 3** werden verschillende niet-selectieve en nieuw ontwikkelde matig selectieve aanrijkingsmedia geëvalueerd op hun resuscitatie capaciteit als monstervoorbereiding voor betrouwbare detectie van *L. monocytogenes* uit verschillende levensmiddelen. In eerste instantie werden de groeiparameters, lag fase (λ) en maximale specifieke groeisnelheid (μ_{max}) bepaald zowel voor gezonde als subletaal beschadigde cellen. Daarna werd een selectie gemaakt van aanrijkingsmedia en werden deze vergeleken met elkaar op hun potentieel om lage aantallen *L. monocytogenes* in natuurlijk gecontamineerde maar ook artificieel gecontamineerde stalen te detecteren binnen de 24 uur. De detectie gebeurde met ofwel bacteriofaag gebaseerde capture (*Listeria* Capture kit, Profos, Regensburg, Germany) gecombineerd met uitplatingen op een selectief medium of door gebruik te maken van fluorescentie *in situ* hybridisatie (FISH) met de VIT-*Listeria* kit (Vermicon, Munich, Germany). De kinetiek van de resuscitatie en de groei van *L. monocytogenes* in de verschillende aanrijkingsmedia toonden dat detectie van lage aantallen subletaal beschadigde cellen minstens een overnacht aanrijking nodig had. Het gebruik van een niet-selectief medium toonde veelbelovend resulaten voor *L. monocytogenes in vitro* maar bleek geen optie voor de detectie van *L. monocytogenes* te zijn uit levensmiddelen (*in situ*). De nood aan een gematigd selectief medium was duidelijk. De studie toonde aan dat gebruikmakend van de juiste combinatie van gepaste aanrijking en een vooruitstrevende detectie techniek, detectie van *L. monocytogenes* binnen 24 uur mogelijk was.

In **Hoofdstuk 4** werd er eveneens een vergelijkende studie gemaakt van λ en μ_{max} van gezonde, gestresseerde en subletaal beschadigde *E. coli* O157 cellen in tien verschillende aanrijkingsmedia. Zowel *in vitro* als *in situ* experimenten werden uitgevoerd en voor de *in situ* experimenten werden er verschillende detectie methoden gebruikt. Twee real-time PCR methoden, namelijk de iQ-Check^{TM} *E. coli* O157:H7 PCR kit (Bio-Rad, Nazareth Eke, Belgium) en de TaqMan® *E. coli* O157:H7 PCR kit (Applied Biosystems, Lennik, Belgium) werden gebruikt. Er werd eveneens gebruik gemaakt van twee capture/uitplaat methoden, de *E. coli* O157 Anti-O157 Dynabeads (Dynal, Oslo, Norway) en de *E. coli* O157 bacteriofaag gebaseerde capture Kit (Profos). In deze studie werd er aangetoond dat TSB (3) zorgde voor de snelste groei van *E. coli* O157 (μ_{max} = 1.00 ± 0.06 u^{-1}) maar ook dat dit aanrijkingsmedium er niet in slaagde om de oxidatief gestresseerde *E. coli* O157 te resusciteren. TSB (1), TSBYE, TSB met 8 mg/l novobiocine + 16 mg/l vancomycine (TSB+), BPW en BPW met 8 mg/l vancomycine (BPW+V) zorgden onafhankelijk van de voorgeschiedenis van de cellen voor resuscitatie van de gestresseerde en subletaal beschadigde cellen. mTSB+N, EC, ECred, TSB (3) and TSB met 8 mg/l novobiocine + 16 mg/l vancomycine + 2 mg/l rifampycine + 1 mg/l K-telluriet + 1.5 g/l Bile salts *N°3* (TSB++) echter faalden allen minstens voor één stress om de cellen te herstellen. TSB (1), TSB+, BPW en BPW+V werden uit de *in vitro* experimenten geselecteerd op basis van een korte λ en een snelle μ_{max}. Hun capaciteit, om op een betrouwbare manier, lage aantalen van gezonde, gestresseerde en subletaal beschadigde cellen te resusciteren uit verschillende levensmiddelen (soja scheuten, gefermenteerde worst, rauwe melk en gehakt rundvlees) werd vergeleken. mTSB+N werd eveneens meegenomen in deze vergelijkende studie als referentie medium. Bij een lage artificiële contaminatie (ongeveer 1 KVE/g) van de levensmiddelen met gezonde cellen, slaagden zowel BPW/BPW+V, TSB, TSB+ en mTSB+N erin om de cellen voldoende aan te rijken voor detectie na 6 uur bij 41.5°C. De studie toonde echter ook aan dat mTSB+N, het aanrijkingsmedium vooropgesteld door de ISO 16654:2001, er niet in slaagde om de subletaal beschadigde cellen te resusciteren, terwijl BPW/BPW+V daar wel in slaagde. Deze studie toonde aan dat bij gebruikt van mTSB+N bij de aanwezigheid van lage aantallen subletaal beschadigde cellen, vals negatieve resultaten bekomen kunnen worden.

In **Hoofdstuk 5** werd een vergelijkende studie gemaakt van Bolton en Preston als aanrijkingsmedia voor een betrouwbare detectie van *C. jejuni* (zowel gezonde als ingevroren cellen). Als reincultuur vertoonde *C. jejuni* een snellere resuscitatie en groei in het Bolton medium in vergelijking met het Preston medium. Wanneer *C. jejuni* samen met extended-spectrum-beta-lactamase (ESBL) producerende *Escherichia coli* geïsoleerd uit gehakt gevogeltevlees van België, geincubeerd werd, domineerde de *E. coli* groei in het Bolton medium en overgroeiden deze *E. coli* de mCCDA isolatieplaten. Dit resulteerde in het feit dat *C. jejuni* niet kon gedetecteerd worden met de ISO 10272:2006 standaard methode. Het Preston aanrijkingmedium vertoonde een tragere groei van de *C. jejuni* cellen in vergelijking met het Bolton medium, maar de geïsoleerde ESBL *E. coli* waren niet in staat tot groei in het Preston aanrijkingsmedium. Het gebruik van eenzelfde antibioticum (sodium cefoperazone) zowel in de aanrijkings- als in de isolatiefase zou een verklaring kunnen zijn voor het optreden van de eerder vastgestelde problemen voor detectie van *C. jejuni* volgens de ISO methode. De hoge aantallen ESBL *E. coli* aangerijkt in het Bolton medium verbergen de *C. jejuni* cellen op het isolatiemedium. Het gebruik van een real-time PCR methode voor de detectie van *C. jejuni* omzeilde de problemen die optraden op de mCCDA platen en kon er voor zorgen dan *C. jejuni* werd gedetecteerd na 24 aanrijking in Bolton. Dit zowel voor de gezonde als de ingevroren *C. jejuni* cellen.

In **Hoofdstuk 6,** werd er een vergelijkende studie gepresenteerd over verschillende aanrijkingsmedia en DNA extractie methoden om lage aantallen van gezonde maar tevens subletaal beschadigde *Salmonella enterica* cellen uit chocolade te detecteren met real-time PCR. De real-time PCR kit iQ-Check™ *Salmonella* II (Bio-Rad), werd gebruikt als detectiemethode. VIDAS® Easy SLM (BioMérieux, Marcy l'Etoile, France) en ISO 6579:2005 (International Standards Organization, 2005b) werden beide gebruikt om de real-time PCR resultaten te verifiëren. Beide alternatieve methoden (VIDAS® en PCR) werden eerder gevalideerd door AFNOR als toepasbaar voor alle levensmiddelen. In de studie voorgeteld in dit hoofdstuk werd specifiek het levensmiddel chocolade onder de loep genomen. Twee DNA extractie protocollen, Standard en Easy extraction protocol beide bruikbaar voor de toepassing van iQ-Check™ *Salmonella* II kit, werden vergeleken. Het Easy extraction protocol toonde de beste resultaten wanneer chocolade geanalyseerd werd. Gebufferd pepton water, het vooraanrijkingsmedium vooropgesteld door ISO 6579:2005 slaagde er niet in om de subletaal beschadigde cellen voldoende aan te rijken voor detectie. Ook werd er in deze studie aangetoond dat de iQ-Check *Salmonella* II bruikbaar was als detectie methode voor natuurlijk met *S. enterica* gecontamineerde chocolade, indien de juiste monstervoorbereidingsstappen toegepast werden. Alle positieve PCR resultaten konden bevestigd worden door gebruik te maken van ISO 6579:2005 of VIDAS® Easy SLM daar minstens 250 KVE/ml *S. enterica* nodig waren om een positief PCR signaal te genereren in chocolade met de gebruikte technieken. In deze studie werd er aangetoond dat het nog steeds een uitdaging is om *S. enterica* cellen uit chocolade te detecteren met gebruik van real-time PCR als detectie methoden door de aanwezigheid van enerzijds lage aantallen subletaal beschadigde cellen, maar anderzijds ook door de inhibitie van de PCR reactie.

De kennis verworven in bovenstaande hoofdstukken heeft geleid tot de formulering van een algemene discussie, voorgesteld in **Hoofdstuk 7,** over de problemen betreffende resuscitatie en ophoping van de betrokken pathogenen. Eveneens werden er een aantal selectie criteria voor snelle microbiologische analyse technieken besproken. De conclusies uit dit doctoraatsproefschrift en de resterende hekelpunten werden aangekaart. Ook werden de essentie en de toegevoegde waarden van dit proefschrift samengevat. Uiteindelijk leidde dit tot de geformuleerde perspectieven voor verder onderzoek.

Chapter 1

Food safety and pathogen detection: literature overview

Partially redrafted after

Vicky Jasson, Liesbeth Jacxsens, Pieternel Luning, Andreja Rajkovic and Mieke Uyttendaele

(2009)

Food Microbiology, submitted

Chapter 1 Food safety and pathogen detection: literature overview

As the consumer becomes more aware of his health the focus is set on a more balanced diet and the link is made with healthy foods. Healthy foods are considered to be foods that are fresh, with presence of vitamins, proteins and fibres, but low contents of sugar, fat and salt. Also preservatives are often perceived to be bad and therefore unwanted in these “healthy foods”. Besides the concern for his health, the consumer often prefers ready-to-eats foods as this type of products are fitting into our busy way of life. These two facts undeniably result in high demands on our foods regarding microbial food safety.

Table 1.1. General characteristics of the “big 4” food-borne pathogens. Data based on EFSA (EFSA, 2009)

Characteristic	*Campylobacter* spp.	*Salmonella* spp.	*L. monocytogenes*	*E. coli* O157:H7
Reported confirmed cases in EU	200507	151995	1554	1551
Disease incidence	Frequent	Frequent	Rare, but with high fatality rate (20%)	Rare, but with severe complications (HUS*)
Most important food	Poultry	Poultry, pork meat and eggs [(a)]	Ready-to-eat foods with extended shelf-life	Ground beef and sprouted seeds [(b)]
Disease	Campylobacteriosis	Salmonellosis	Listeriosis	Hemorrhagic colitis
Typical clinical symptoms	Fever, headache, muscle pain, nausea, malaise followed by diarrhea	Acute onset of fever, abdominal pain, diarrhea, nausea and sometimes vomiting	Mild flu-like symptoms and diarrhea	Mild to bloody diarrhea, abdominal cramps and usually without fever
More severe clinical symptom	Guillain-Barré syndrome	Dehydratation and bloodstream infection	Septicaemia and meningitis-encephalitis to abortion and dead	HUS* and TTP**
Ages of most reported confirmed cases	0-4 and 15-24 years	0-4 years [(c)]	0-4 and ≥ 65 years	0-4 years [(d)]
Most important sero- or genotypes	*C. jejuni* and *C. coli*	*S. enterica* subsp. *enterica serotype Typhimurium and serotype Enteritidis*	Serotypes 1/2a, 1/2b, and 4b	O157:H7 [(e)]

***: haemolytic ureamic syndrome**
****: thrombotic thrombocytopenic purpura**
(a), (b), (c), (d) and (e): The table shows a simplified summary of information which is described more in detailed in section 1.1 of this PhD.
(a): also water or cross-contamination of fruits and vegetables by water and low moisture products
(b): also water or cross-contamination of fruits and vegetables by water and dry or acid products
(c): all healthy people
(d): also ≥ 65 years
(e): also *E. coli* serotype O26, O145, O111 and O103 are important regarding HC and HUS

Food-borne illnesses are unacceptable as they cause lots of human pain and economical losses. Microorganisms, associated with food-borne illness, most often in the spotlight are *Listeria monocytogenes*, *Campylobacter* spp., *Salmonella* spp. and *Escherichia coli* O157:H7 (Table 1.1), also often referred to as the "big 4". Other microorganisms such as viruses (especially Noroviruses) are not targeted yet in routine analysis or surveillance studies due to the lack of appropriate reference methods. Reliable and fast detection techniques are needed to ensure the consumer that fresh and healthy food is also safe. Some of the general characteristics of the "big 4" are summarized below in the order of their frequency of causing illness.

1.1 Food-borne pathogens

1.1.1 *Campylobacter* spp.

1.1.1.1 General characteristics

The genus *Campylobacter* compromises 17 species: *C. fetus*, *C. hyointestinalis*, *C. sputorum*, *C. mucosalis*, *C. concisus*, *C. curvus*, *C. rectus*, *C. gracilis*, *C. showae*, *C. jejuni*, *C. coli*, *C. lari*, *C. upsaliensis*, *C. canadensis*, *C. insulaenigrae*, *C. lanienae* and *C. helveticus*. *Campylobacter* spp. are Gram negative non-spore forming bacteria and cells are mostly slender curved, spiral-shaped rods of 0.2 to 0.8 µm wide and 0.5 to 5 µm long. Cells of most species are motile with corkscrew-like motion by means of a single-polar flagellum at one or both ends. They are catalase and oxidase (except *C. gracilis*) positive and urease negative. The optimal atmosphere for growth is 5% O_2, 10% CO_2 en 85% N_2, and is called micro-aerobic atmosphere. *C. jejuni*, *C. coli*, *C. lari* and *C. upsaliensis* do not grow below 30°C and are therefore characterized to be thermo-tolerant. *Campylobacter* spp. are fastidious organisms (Nachamkin *et al.*, 2008).

1.1.1.2 Significance of *Campylobacter* spp.

The most important human enteropathogens of the *Campylobacter* genus are *C. jejuni* and *C. coli* followed by *C. lari* and *C. upsaliensis*. Within *C. jejuni* two sub-species are recognized: *C. jejuni* subsp. *jejuni* and *C. jejuni* subsp. *doylei*, but the pathogenesis of *C. jejuni* subsp. *doylei* is still unknown (Nachamkin *et al.*, 2008). Food poisoning caused by *Campylobacter* spp. is called campylobacteriosis. Fever, headache, muscle pain, and malaise are present 12 to 48 h before onset of diarrhea. The diarrhea is often accompanied by cramping and abdominal pain. The abdominal pain can be severe enough to mimic appendicitis. The pattern of diarrhea varies from loose stools to profuse, bloody, slimy and/or foul-smelling stools. Generally, these symptoms are self-limiting and may persist from one day to a week or longer, but infection is rarely life-threatening. No antibiotics are used against the infection (Smith, 1995). Campylobacteriosis has been linked with subsequent development of Guillain-Barré syndrome (GBS), which usually develops two to three weeks after the initial illness. Some strains of *Campylobacter* spp. produce a cholera-like enterotoxin, which is important in the watery diarrhea observed in infections (Levin, 2007). In a small number of cases, the infection may be associated with haemolytic uraemic syndrome (HUS) and thrombotic thrombocytopentic purpura through a poorly understood mechanism (Kamioka *et al.*, 2007).

During 2007, campylobacteriosis was the most frequently reported zoonotic disease in humans in the European Union with 200507 confirmed illnesses (EFSA, 2009). The most susceptible groups of people are between 0-4 and 15-24 years old. Reason for this can be found in immunological response (0-4 years) and eating habits (15-24 years). Immunity against *Campylobacter* spp. is gained over the years with unintentional indigestions of cells (Altekruse *et al.*, 1999; Blaser *et al.*, 1987).

1.1.1.3 <u>Transmission routes of *Campylobacter* spp.</u>

C. jejuni and *C. coli* are normally found in cattle, swine, and birds, where they are non-pathogenic. *C. upsaliensis* is often found in cats and dogs and *C. lari* in particular in seabirds. The common routes of transmission to humans are 1) ingestion of contaminated food 2) drinking water and 3) recreation water. Contact with contaminated poultry, livestock, or household pets, especially puppies, are also transmission routes, but animals farmed for meat are the main source of campylobacteriosis. Due to the high level of consumption the risk of getting campylobacteriosis is the highest from eating chicken (Humphrey *et al.*, 2007; Vellinga and Van Loock, 2002).

In foodstuffs, the highest proportion of *Campylobacter* spp. positive samples were reported for fresh poultry meat in Europe (26% of samples were found positive). Poultry meat is still the most important food-borne source of *Campylobacter* spp. as the occurrence of the bacteria remains at high levels throughout the food chain, from live animals to meat retail level. In other foodstuffs such as unpasteurized milk, only occasionally *Campylobacter* spp. was detected (EFSA, 2009). A Belgium-wide survey of *Campylobacter* spp. contamination in chicken meat preparation revealed that 48.02% of these food samples were contaminated (Habib *et al.*, 2008a).

1.1.2 *Salmonella* spp.

1.1.2.1 General characteristics

Salmonella spp. are members of the *Enterobacteriaceae* family. They are facultative anaerobic Gram negative rods (0.7-1.5 µm wide and 2.0-5.0 µm long), non-spore forming, which are usually motile with peritrichous flagella. They are catalase positive, but oxidase and urease negative. The genus *Salmonella* contains two species: *Salmonella enterica* and *Salmonella bongori*. *Salmonella enterica* is an important agent of food-borne illness. This species is sub-classified into 6 sub-species (*S. enterica* subsp. *enterica, S. enterica* subsp. *salamae, S. enterica* subsp. *arizonae, S. enterica* subsp. *diarizonae, S. enterica* subsp. *boutenae and S. enterica* subsp. *indica*) of which *S. enterica* subsp. *enterica* is the most important regarding human illness. The genus *Salmonella* can be subdivided into more than 2400 serotypes. *Salmonella enterica* subsp. *enterica* serotype Typhimurium (*S.* Typhimurium) and *Salmonella enterica* subsp. *enterica* serotype Enteritidis (*S.* Enteritidis) are the most frequently isolated serotypes. Some serotypes are further sub-divided by their sensitivity towards bacteriophages (phage types), antibiotics or heavy metals, by biochemical characteristics (biovars or biotypes) or their sensitivity to or production of bacteriocins (Bell and Kyriakides, 2002).

1.1.2.2 Significance of *Salmonella* spp.

Human salmonellosis is usually characterised by the acute onset of fever, abdominal pain, diarrhea, nausea, and sometimes vomiting. Symptoms are often mild and most infections are self-limiting, lasting a few days. However, in some patients, the infection may be more serious and the associated dehydration can be life-threatening. In these cases, as well as when *Salmonella* spp. causes bloodstream infection, effective antibiotics are essential for treatment. Salmonellosis has also been associated with long-term and sometimes chronic sequelae e.g. reactive arthritis (Chiu *et al.*, 2004).

Salmonellosis was responsible for approximately 151995 illnesses in the European Union in 2007 (EFSA, 2009). *Salmonella* spp. are the second most common cause of bacterial food-borne illness. Especially young children (age 0-4) are vulnerable to *Salmonella* infection (EFSA, 2009; Graham, 2002) in particular infants (Marcus, 2008).

1.1.2.3 Transmission routes of *Salmonella* spp.

Salmonella spp. are normally found in different domestic and wild animals including birds. In animal populations, *Salmonella* spp. is most frequently detected in poultry flocks (EFSA, 2009). The common routes of transmission to humans are 1) ingestion of contaminated food and 2) direct contact with infected animals and humans. Some animals and people are infected without being ill at all. They are known as "carriers" and can still spread the disease to others. Because the intestinal tract of a wide range of domestic and wild animals is the common reservoir for *Salmonella* spp., a variety of foodstuffs covering both food of animal (animal reservoir) and plant origin (cross-contamination) can be potential sources of infections. Transmission to humans mostly occurs when these microorganisms are introduced in food production or preparation areas. Inadequate storage temperatures can allow growth of *Salmonella* spp., but also inadequate cooking or cross contamination to ready-to-eat food can cause infection by this pathogen. In addition it is important to notice that although *Salmonella* spp. do not form spores they can survive for long time especially in foods with low water activity (a_w).

Many foods (EFSA, 2009) have been identified as vehicles for the transmission to humans in particular meat (Urfer *et al.*, 2000), poultry, eggs (Gantois *et al.*, 2009), milk (Brouard *et al.*, 2007), chocolate (Werber *et al.*, 2005), fruit and vegetables (Little and Gillespie, 2008). Human *S.* Enteritidis cases are most commonly associated with the consumption of contaminated eggs and poultry meat, while *S.* Typhimurium cases are mostly associated with the consumption of contaminated pork, but also poultry and beef meat.

Salmonella was most often found in fresh poultry and pork meat where proportions of positive samples, on average 5.5% and 1.1%, were detected respectively. 0.8% of table eggs were found positive with *Salmonella* spp., while dairy products, vegetables and fruit were rarely found to contain the bacterium (EFSA, 2009).

1.1.3 *Listeria monocytogenes*

1.1.3.1 General characteristics

The genus *Listeria* consists of six species: *L. monocytogenes*, *L. innocua*, *L. ivanovii*, *L. welshimeri*, *L. seeligeri* and *L. grayi*. *L. monocytogenes* is a Gram positive, non-spore forming, facultative anaerobic, rod shaped bacterium (0.5 µm wide and 1-2 µm long). It is catalase positive, oxidase and urease negative and expresses a β-haemolysin which causes destruction of red blood cells. When viewed with light microscopy *L. monocytogenes* exhibits characteristic tumbling motility. Although *L. monocytogenes* is actively motile by means of peritrichous flagella at room temperature (20-25°C), the organism does not synthesize flagella at body temperatures (37°C) (Ryser and Marth, 1999).

1.1.3.2 Significance of *L. monocytogenes*

Listeriosis is the disease caused by bacteria from the genus *Listeria*, but *L. monocytogenes* is the major pathogenic species in both animals and humans. Some cases of human infection caused by *L. ivanovii* and *L. seeligeri* have been reported. *L. ivanovii* infections may account for a significant proportion of cases of listeriosis in domestic animals and there is some evidence for rare infection caused by *L. innocua* in domestic animals. The species *L. welshimeri* and *L. grayi* have not been reported to cause disease. There are 13 serotypes of *L. monocytogenes* which can cause disease. More than 90% of the human isolates belong to only three serotypes namely 1/2a, 1/2b, and 4b (McLauchlin *et al.*, 2004). Especially young, old, pregnant and immunocompromised people, abbreviated as the YOPI group, are vulnerable to *Listeria* infection and severe illness mainly occurs in this group. Symptoms vary, ranging from mild flu-like symptoms and diarrhea to life-threatening infections characterised by septicaemia and meningitis-encephalitis. In pregnant women the infection can spread to the foetus, which may either be born severely ill or die in the uterus and results in abortion (Farber and Peterkin, 1991).

L. monocytogenes is considered to be one of the most virulent food-borne pathogens with a high fatality rate (20%) resulting in the leading cause of deaths among food-borne bacterial pathogens. On average, *L. monocytogenes* was the most severe pathogen associated with outbreaks in 2007 with 1554 confirmed cases of listeriosis in the Europeans Union (EFSA, 2009). Although human infections are rare importance is given to *L. monocytogenes* due to the high associated mortality rate.

1.1.3.3 Transmission routes of *L. monocytogenes*

Listeria spp. are ubiquitous organisms that are widely distributed in the environment, especially in plant matter and soil. The principal reservoirs of *Listeria* spp. are soil and water. Other reservoirs include infected domestic and wild animals. The main route of transmission to both humans and animals is believed to be through consumption of contaminated food or feed (Farber and Peterkin, 1991).

L. monocytogenes has been associated with foods such as raw milk, pasteurized fluid milk, cheeses (particularly soft-ripened varieties) and ice cream (Kozak *et al.*, 1996), raw vegetables, fermented sausages, raw and cooked poultry (Waldroup, 1996), raw meats (Thevenot *et al.*, 2006; Waldroup, 1996; Zhu *et al.*, 2005), processed meat with extended shelf-life and raw and smoked fish (Novotny *et al.*, 2004). It is considered to be one of the most important post-contamination pathogens. Heating kills *L. monocytogenes*, but the pathogen is capable to grow at temperatures as low as 0°C resulting in a risk of multiplication (although slow) in refrigerated foods which makes the occurrence in ready-to-eat foods (RTE) with a relatively long shelf-life of particular concern (Bayles *et al.*, 1996; Doyle *et al.*, 2001). EFSA reported that *L. monocytogenes* was seldom detected above the legal safety limit on RTE foods, whereas numbers exceeding this limit were most often found in smoked fish and other RTE fishery products and in lesser extent by RTE meat products and cheeses (EFSA, 2009).

1.1.4 *Escherichia coli* O157:H7

1.1.4.1 General characteristics

Escherichia coli O157:H7 is one of the hundreds of serotypes of the bacterium *Escherichia coli*. *E. coli* is a Gram negative bacterium which is facultative anaerobic, generally motile, slim, occasionally slightly curved short rods ranging in size from 1-5 µm in length and 0.3-0.4 µm in diameter (Yoon and Hovde, 2008).

While most *E. coli* strains are harmless and normally found in the intestines of warm blood animals, the O157:H7 serotype is a member of the pathogenic *E. coli* known as enterohemorrhagic *Escherichia coli* or EHEC. The letter "O" in the name refers to the somatic antigen number, whereas the "H" refers to the flagella antigen (LeBlanc, 2003). *E. coli* O157:H7 colonization of the intestinal epithelial surface is accompanied by a characteristic attaching and effacing (A/E) histopathology marked by localized destruction of brush border microvilli, intimate attachment to enterocytes and rearrangement of cytoskeletal components beneath adherent bacteria. *E. coli* O157 may produce Shiga-like toxins also called Verocytotoxins. These toxins cause the severity of the illness. *E. coli* O157:H7 are sometimes also called Verocytotoxin producing *E. coli* (VTEC) or Shiga-like Toxin producing *E. coli* (STEC) referring to their toxin production (Welinder-Olsson and Kaijser, 2005).

E. coli O157:H7 normally cannot ferment sorbitol within 24 h, cannot hydrolyze 4-methylumbrelliferyl-D-glucuronidase (MUG) and is unable to grow at temperatures higher than 44.5°C in the presence of 0.15% bile salts. These characteristics can be used to differentiate between *E. coli* O157:H7 from other *E. coli* (Beneduce *et al.*, 2003).

1.1.4.2 Significance of *E. coli* O157:H7

The disease caused by *E. coli* O157:H7 is called hemorrhagic colitis (HC) but is also known as the "hamburger disease" because of its association with eating hamburgers. This disease is rather sporadic (Bell *et al.*, 1994). The symptoms associated with *E. coli* O157:H7 infections in humans vary from mild to bloody diarrhea, which is often accompanied by abdominal cramps, usually without fever, but infections can result in the serious disease called haemolytic uraemic syndrome (HUS). HUS is characterised by acute renal failure, anaemia and lowered platelet counts. HUS develops in up to 10-15% of patients infected with *E. coli* O157:H7 and is the leading cause of acute renal failure in young children (age 0-4) (Amirlak and Amirlak, 2006). *E. coli* O157:H7 has also been associated with the development of thrombotic thrombocytopenic purpura (TTP) which is together with HUS a life-threatening complication (LeBlanc, 2003).

A large number of serotypes of *E. coli* have been recognised as verocytotoxin (VT) producers. However the O157 serotype (VTEC O157) is the one most frequently reported to be associated with the human disease. Other important VTEC serotypes include: O26, O103, O111 and O145 because these serotypes have been increasingly isolated from clinical cases and outbreaks as well as from animals and environmental sources (Posse *et al.*, 2008). VTEC infections were responsible for 2904 confirmed illnesses in the European Union in 2007 (EFSA, 2009), 1571 (54.1%) cases were reported to be caused by O157 serotype. In total, 103 HUS cases were associated with these 2904 VTEC infection (EFSA, 2009).

1.1.4.3 Transmission routes of *E. coli* O157:H7

E. coli O157:H7 is found in the intestines of healthy ruminants. The toxin requires highly specific receptors on the cells' surface in order to attach and enter the cell. Species such as cattle, swine, and deer which do not carry these receptors may harbor these pathogenic bacteria without any symptoms of illness, shedding them in their feces from which they may be spread to humans. Water-borne transmission occurs through swimming in contaminated lakes, pools, or drinking inadequately treated water. The organism is, due to its low infection dose, easily transmitted from person-to-person and has been difficult to control e.g. in child day-care centers (Karch *et al.*, 2005).

Meat can become contaminated during slaughter and organisms can be thoroughly mixed into beef when it is ground into hamburger, resulting in the major source of infection. Other sources include consumption of unpasteurized milk and apple juice, raw sprouted seeds, lettuce, salami and contact with infected live animals (Erickson and Doyle, 2007; McClure and Hall, 2000). Among animals and foodstuffs, VTEC is most often reported in cattle and bovine meat. The bacterium was very rarely recovered from vegetables (EFSA, 2009).

1.2 Detection methods for pathogens in food

Minimizing the risk of food-borne illness is a complete from farm-to-fork approach (including the consumer), based on good manufacturing practice (GMP) and Hazard Analysis Critical Control Points (HACCP), but also on appropriate sampling and detection methods. Despite this farm-to-fork approach, authorized control is only possible until the level of retail. Detection methods for pathogens in foods are the ultimate (final) tool applicable to control/check the food for microbiological hazards before it arrives at the consumer. It is therefore of the utmost importance that these techniques are reliable, appropriately applied and also fast in case of the need for rapid action.

1.2.1 Definition of methods

Methods for microbial analysis comprise methods for enumeration (quantitative methods) and methods for detection (qualitative methods). In many occasions, the classical (conventional) culture methods (based upon multiplication of target organisms in broth or agar media to numbers visual by the naked eye as turbidity, colour change or colonies) are referred to as the reference methods for microbial analysis. However, also other types of methods based upon other principles of detection (e.g. molecular based methods) may be recognized or acknowledged as reference methods if they are perceived as providing a more accurate result. In microbiology, reference methods are most often open access methods (not proprietary methods) and have been subjected to some kind of collaborative study or peer review by independent labs to gain a wide acceptance of proven performance.

For official analysis, most often reference methods are recommended, however, alternative methods, usually proprietary methods can be accepted as well if they have been shown to provide equivalent results by means of a validation study. In particular cases, e.g. for emerging pathogens in foods such as Norovirus, VTEC, etc. for which international agreement yet has to be obtained, in-house validated methods may be used for official analysis.

In the frame of assurance of food safety, often proprietary methods are used, but still classical reference methods are used (to some extent automated or with minor modifications) and occasionally in-house methods. Definitions of these various types of methods are depicted in Table 1.2.

Table 1.2. Definition of type of methods available for microbial analysis of foods.

Type of Method	Definitions according to Working document of revised ISO 16140 (2009) Method Validation – Part 1 : Terminology
Methods suitable for official controls *(as defined by EC 882/2004)*	Method acknowledged for use in "official control"; official control meaning any form of control that the competent authority performs for the verification of compliance with legislation. *Official control methods shall comply with internationally recognized methods such as CEN methods or methods agreed upon in national bodies. Or in absence of these (inter)national methods other methods shall be used fit for intended purpose and developed according to scientific protocols.*
Reference method	An Internationally recognized method and widely accepted. *A reference method has been thoroughly investigated and clearly and exactly describes the necessary conditions and procedures, for the measurement of one or more property values. Accuracy and precision should be commensurate with its intended use and can therefore be used to assess the accuracy of other methods for the same measurement. As a rule, the reference method is a standard or a commonly used method taken as reference.*
Alternative method	Method of analysis for detection or estimation, of the same analyte as is measured by using the corresponding reference method, for a given category of products. *The method can be proprietary or non-commercial and covers an entire analytical procedure, that is, from the preparation of samples to the test results either as such or may include references to other procedures in order to be complete* *The alternative method exhibits attributes appropriate to the user' needs, for example:* - *speed of analysis and/or response* - *ease of execution and/or automation* - *analytical properties (precision, accuracy, limit of detection, etc.)* - *miniaturisation* - *reduction of cost*
Proprietary method	Method with a registered trademark/brand name, which is owned and generally marketed by a commercial company. *Generally, some of the components of the method are undisclosed.*
Qualitative method	Method of analysis that detects either the presence or absence of the analyte, either directly or indirectly, in a certain quantity of the sample.
Quantitative method	Method of analysis whose response is the amount of the analyte measured either directly (e.g. enumeration in a mass or volume), or indirectly (e.g. colour absorbance, impedance, etc.) in a certain quantity of sample.
Validated method	An alternative method that has demonstrated based on the validation study to provide adequate confidence that the results obtained by this method are comparable to those obtained by the reference method.

1.2.2 Reference methods for detection

Detection methods for the "big 4", *Campylobacter* spp., *Salmonella* spp., *L. monocytogenes* and *E. coli* O157, being the subject of surveillance programs and intensive testing in the frame of verification and validation of food safety management systems (FSMS), normally employ enrichment-plating techniques. As virtually all pathogens in foods are sub-lethally injured (Wu, 2008), the detection method encompasses often a two-step enrichment procedure: 1) a resuscitation or pre-enrichment step lasting between a few hours to overnight incubation in a non or half selective medium to enable recovery and limited outgrowth (10^2 to 10^4 CFU/ml) of the stressed target organism; 2) a period of enrichment in a selective medium to suppress the competitive flora and enabling multiplication of the target organism to attain high enough levels (approximately 10^4 to 10^6 CFU/ml) for subsequent isolation or detection by a variety of techniques.

In the classical culture methods, after enrichment the target organism is isolated on a selective differential agar medium. Finally, purification of presumptive typical colonies and confirmation is performed using a defined number of morphological, biochemical, physiological and sometimes serological (agglutination) tests. These confirmation tests aim to verify a restricted number of typical characteristics which are easily determinable and enable good differentiation between competitive closely related strains.

Standard methods for *Campylobacter* spp. (International Standards Organization, 2006b), *Salmonella* spp. (International Standards Organization, 2005b), *L. monocytogenes* (International Standards Organization, 1996) and *E. coli* O157 (International Standards Organization, 2001a) are exemplified in Figures 1.1, 1.2, 1.3 and 1.4 respectively. These reference methods do not differ a lot from the conventional detection methods, developed more the 100 years ago (Feng, 1996; Fung, 1994). Overall the time to obtain a negative test result (absence of typical colonies on the isolation medium) or a presumptive positive results (presence of typical colonies on the isolation medium) is approximately 1 to 4 days, whereas a fully confirmed positive test results may take up to 5 to 7 days.

1.2.2.1 *Campylobacter* spp.

A schematic representation of the ISO 10272:2006 (International Standards Organization, 2006b) detection procedure for thermo-tolerant *Campylobacter* spp. from food and feed samples is shown in Figure 1.1. Pre-enrichment and enrichment occur in the same enrichment broth, Bolton broth. Regarding pre-enrichment 37°C for 4 h is implemented to resuscitate the *Campylobacter* spp., while enrichment itself takes place at 41.5°C which provides selectivity.

Bolton broth is designed to aid recovery of injured cells. The base medium contains sodium pyruvate, sodium metabisulphite and haemin to quench toxic compounds that may be formed in the culture medium and to decrease the oxidative stress caused by the medium. The following antibiotics are added to the medium: sodium cefoperazone, which is a broad spectrum antibiotic, vancomycin, which is active against the growth of Gram positive bacteria, trimethoprim lactate especially against *Proteus* spp. and amphotericin B against the growth of yeasts and moulds. Sterile sheep or horse blood or charcoal is added to the media to neutralize the toxic effects of oxygen (Corry *et al.*, 1995). After enrichment, the selective media, modified charcoal cefoperazone deoxycholate agar (mCCDA) and another not defined agar medium are used to isolate colonies. The selective plating medium mCCDA is developed to replace blood with charcoal, ferrous sulphate and sodium pyruvate to counteract the toxic oxygen effects. Sodium deoxycholate is added to inhibit growth of Gram positive bacteria. Selectivity is achieved by cefoperazone and amphotericin B similar to Bolton broth. *Campylobacter* spp. form grey coloured colonies on mCCDA, but suspected colonies have to be confirmed. First there is a microscopic examination of the presumptive colonies which should reveal spiral shaped morphology and motility confirmation. After this morphological confirmation the biochemical confirmations follow. Growth at 25°C in micro-aerobic atmosphere and aerobic growth at 41.5°C should be negative while the oxidase reaction should be positive. Further identification of *Campylobacter* spp. to species level can be performed but is not obligatory (International Standards Organization, 2006b).

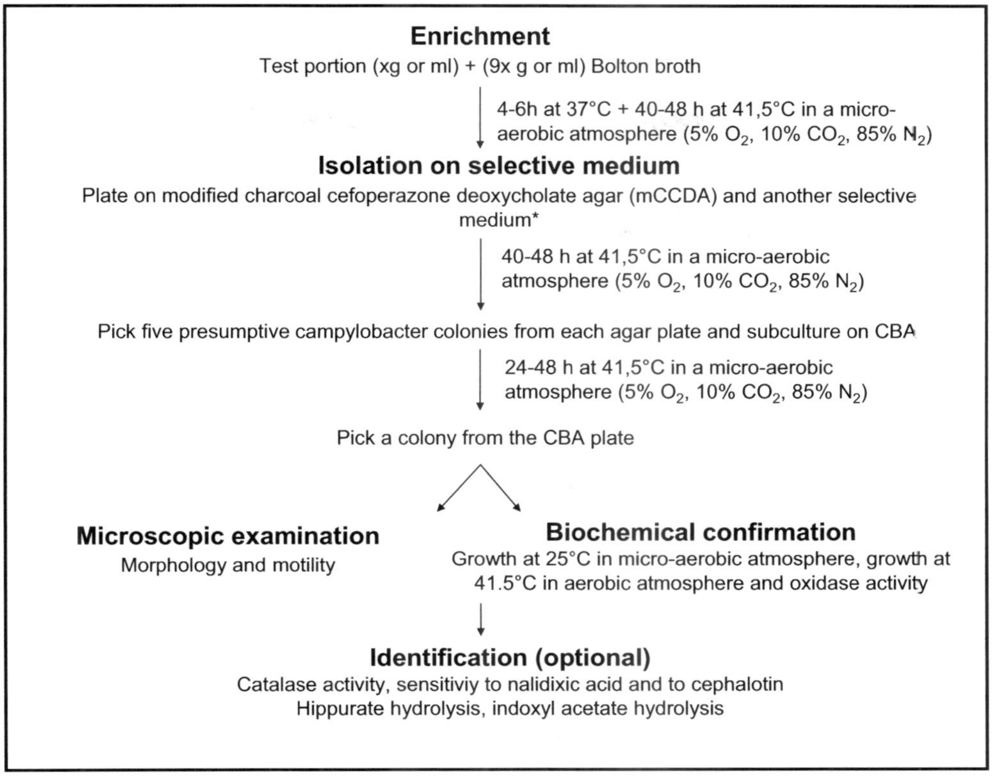

Figure 1.1. Scheme of ISO procedure for the detection of *Campylobacter* spp. from food and feed samples (International Standards Organization, 2006b). *: incubation at its recommended time and temperature profile.

1.2.2.2 *Salmonella* spp.

The schematic representation of the ISO 6579:2005 detection procedure (International Standards Organization, 2005b) for *Salmonella* spp. from food and feed samples is shown in Figure 1.2. Pre-enrichment or resuscitation is performed in Buffered Peptone Water (BPW) at 37°C. BPW is used as enrichment medium as it aids to the recovery of injured *Salmonella* spp. cells and it has the capacity to buffer against pH changes due to growth and metabolism of the microorganisms and those imposed by the food sample (Baylis *et al.*, 2000a). Selective enrichment is performed in respectively Rappaport-Vassiliadis Soya Peptone Broth (RVS) and Muller-Kauffmann Tetrathionate-Novobiocin broth (MKTTn). RVS has the ability to exploit some differentiating characteristics of *Salmonella* spp. compared with other

Enterobacteriaceae. *Salmonellae* have the characteristics to survive at relatively high osmotic pressure and therefore 7.2 g/l NaCl and 13.58 g/l $MgCl_2$ are used in RVS. They are relatively more resistant to malachite green and are able to multiply at relatively low pH values (pH = 5.2) (Atlas, 2006). In MKTTn, ox bile, brilliant green and novobiocin are present to improve selectivity. Brilliant green dye inhibits Gram positive bacteria and the majority of Gram negative bacilli, while novobiocin is added to improve inhibition of *Proteus* spp. (Jeffries, 1959). Organisms which can utilise tetrathionate, such as most *Salmonella,* will flourish.

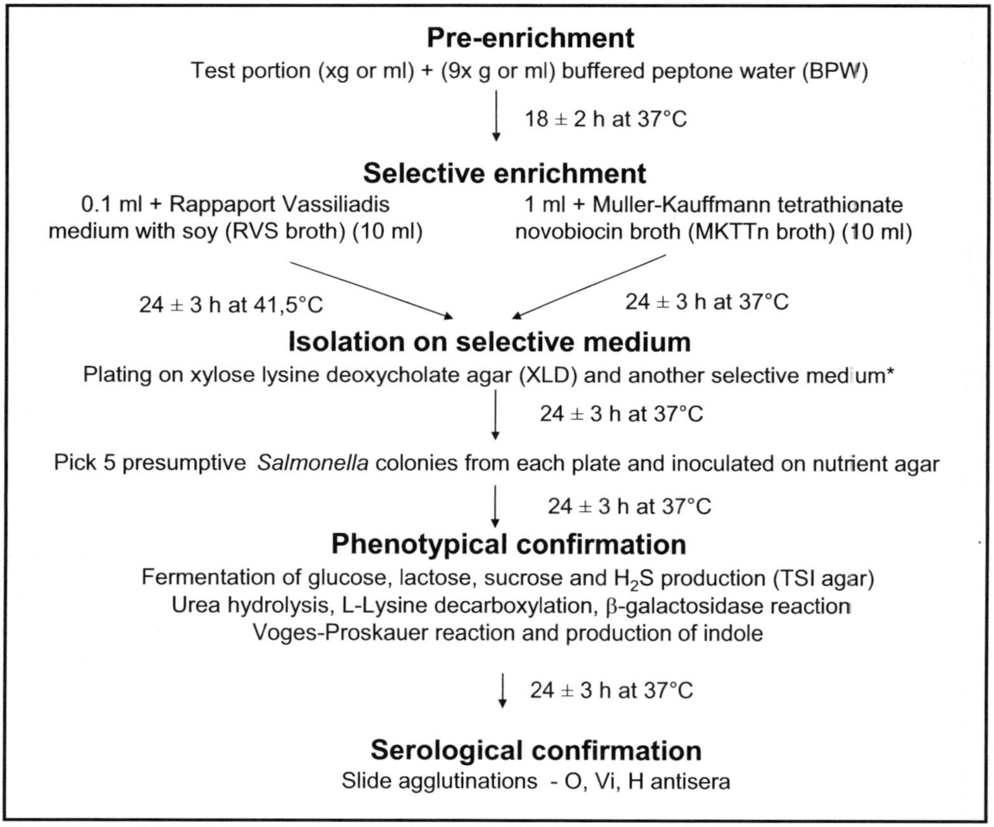

Figure 1.2. Scheme of ISO procedure for the detection of *Salmonella* spp. from food and feed samples (International Standards Organization, 2005b). *: incubation at its recommended time and temperature profile.

After selective enrichment the selective isolation medium xylose lysine deoxycholate (XLD) agar is used to isolate *Salmonella* spp. colonies. XLD relies on xylose fermentation, lysine decarboxylation and production of hydrogen sulphide for the primary differentiation of *Salmonella* spp. Rapid xylose fermentation is almost universal amongst enteric bacteria, but *Salmonella* spp. are differentiated from closely related microorganism due to the inability to ferment lactose or sucrose (Atlas R.M., 2006). After enrichment and isolation, suspected colonies have to be confirmed. Fermentation of glucose, lactose and sucrose and production of H_2S is tested on TSI agar. A typical *Salmonella* spp. isolate would produce acid and gas from the glucose in TSI agar and would not utilize lactose or sucrose. Further it would generate H_2S gas. They are catalase positive but oxidase and urease negative.

1.2.2.3 ***L. monocytogenes***

The ISO 11290:1996 (International Standards Organization, 1996) detection procedure for *Listeria* spp. including the important pathogen *L. monocytogenes* from food and feed samples is shown in Figure 1.3. The pre-enrichment broth is Half Fraser broth which is a modification of the enrichment broth Fraser. Half Fraser, sometimes referred to as Demi or ½ Fraser, contains half of the concentration of nalidixic acid and acriflavine hydrochloride as Fraser broth to aid in the recovery of injured *Listeria* spp. cells. All *Listeria* spp. hydrolyse aesculin to aesculetin which reacts with ferric ions, accompanied with subsequent blackening of the medium. Blackening of the medium is presumptive evidence of the presence of *Listeria* spp. Another advantage regarding the addition of ferric ammonium citrate is that it has been shown that ferric ions enhance the growth of *L. monocytogenes* (Cowart and Foster, 1985). Lithium chloride is included in the medium to inhibit the growth of enterococci as they are also able to hydrolyse aesculin.

After pre-enrichment and enrichment, target colonies are isolated on the Agar *Listeria* Ottaviani & Agosti (ALOA) selective medium and in addition on another selective medium. Again lithium chloride is added combined with a balanced antimicrobial and antifungal mixture to provide selectivity. The incorporation of the chromogenic substrate X-glucoside, for the detection of beta-glucosidase activity, will demonstrate the presence of *Listeria* spp. which appear then as turquoise colonies. The detection of a specific phospholipase C enzyme, produced by the pathogenic *Listeria* spp. *L.*

monocytogenes, *L. seeligeri* and *L. ivanovii*, will distinguish between pathogenic and non-pathogenic *Listeria* spp. This difference is achieved by similar coloured colonies but extra surrounded by a characteristic opaque halo after 24 hours incubation at 37°C due to the breakdown of egg yolk present in the medium (Stessl *et al.*, 2009).

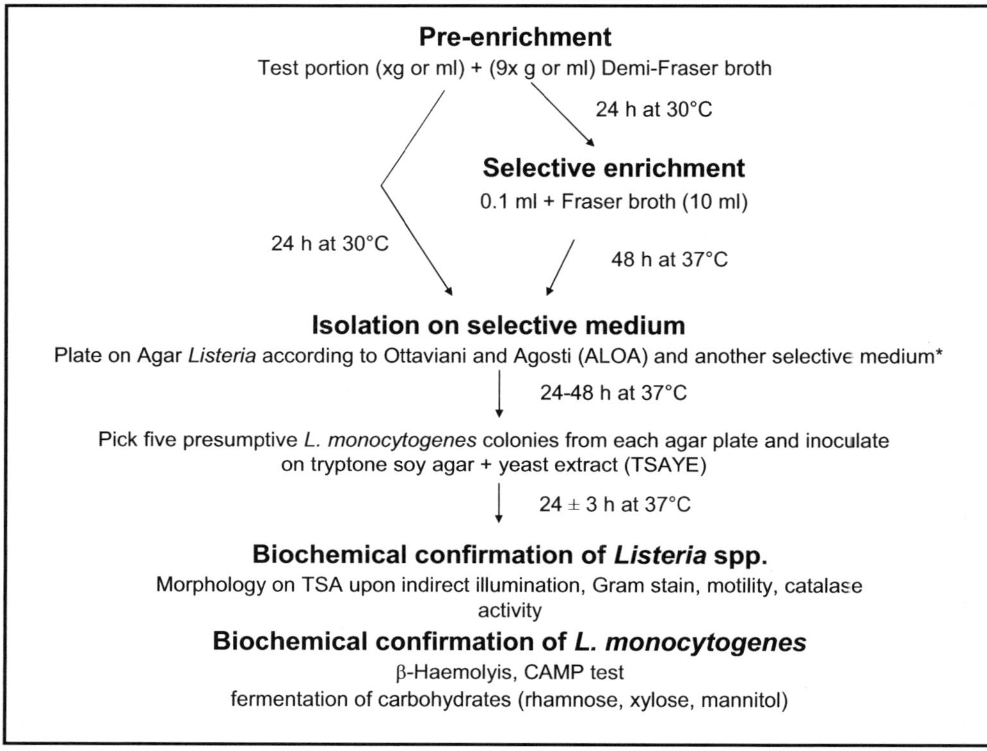

Figure 1.3. Scheme of ISO procedure for the detection of *Listeria* spp. from food and feed samples (International Standards Organization, 1996). *: incubation at its recommended time and temperature profile.

After enrichment and isolation, suspected colonies have to be confirmed. For the confirmation of *Listeria* spp. plating on TSA followed by examination under indirect illumination (Henry light) is performed as *Listeria* spp. can be distinguished by their bluish cast (Lachica, 1990) in combination with Gram staining and motility and catalase activity testing. *L. monocytogenes* can be distinguished from other *Listeria* spp. by their β-haemolysis capacity and the CAMP test (McKellar, 1994). The most important features discriminating between *Listeria* spp. commonly found in food environment are β-haemolysis, the CAMP reaction and xylose fermentation (Beumer and Hazeleger, 2003). Fermentation of xylose can differentiate between *L. monocytogenes* and *L. innocua* and the other *Listeria* spp. More recent methods for the confirmation of suspected colonies are based on DNA-probes and latex agglutination tests (Beumer and Hazeleger, 2003).

1.2.2.4 *E. coli* O157

The ISO 16654:2001 (International Standards Organization, 2001a) detection procedure for the detection of *E. coli* O157:H7 from food and feed samples is shown in Figure 1.4. Modified Tryptone Soya Broth abbreviated as mTSB is used as enrichment broth. The bile salts are added to inhibit the growth of spore formers and enterococci. Novobiocin is added to enhance inhibition of Gram positive organisms. Immunomagnetic separation and concentration is incorporated in the detection protocol because it has proven to aid in sample preparation to eliminate competing flora in the enrichment. As such it can be considered as a secondary enrichment step to fasten up the detection protocol by concentration the target bacteria. Immunomagnetic separation and concentration is performed after 6 hours of enrichment (short enrichment step), and subsequent after 18-24 h of enrichment at 41.5°C (long enrichment step).

Sorbitol MacConkey Agar supplemented with cefixime and K-tellurite (CT-SMAC) is recommended for the isolation of pathogenic *E. coli* O157:H7. The formulation of Sorbitol MacConkey Agar is identical to MacConkey Agar (differential and selective medium to grow Gram negative bacteria that can ferment lactose) except that lactose has been replaced with sorbitol. *E. coli* O157:H7 does not ferment sorbitol and therefore (using neutral red as pH indicator), produces colourless colonies. In contrast, most *E. coli* strains ferment sorbitol and form pink colonies. The level of K-tellurite selects serogroup O157 from other *E. coli* serogroups and inhibits *Providencia* and *Aeromonas* species. Cefixime is inhibitory to *Proteus* spp. (Zadik *et al.*, 1993). Besides CT-SMAC, any other selective plating medium needs to be used in parallel.

Enrichment

Test portion (xg or ml) + (9x g or ml) modified tryptone soya broth containing novobiocin (mTSB+n) pre-heated at 41,5°C

↓ 6 h at 41,5°C subsequent 12-18 h at 41,5°C

Immunomagnetic separation and concentration

-1 ml enrichment
-Immunocapture in microcentrifuge tubes
-Separation with magnetic separator
-Wash steps with modified phosphate buffer
-Resuspend magnetic beads in 100 µl

↓

Isolation on selective medium

Immunomagnetic beads with attached bacteria (50 µl each) on cefixime tellurite sorbitol MaConkey agar (CT-SMAC) and another selective isolation agar*

↓ 18-24 h at 37°C

Pick 5 presumptive *E. coli* O157 colonies from each agar plate inoculated on nutrient agar

↓ 18-24 h at 37°C

Biochemical and serological confirmation

Production of indole and serological agglutinations with *E. coli* O157 antiserum

↓

Genetic characterisation (in reference lab)

VT genes, *eae*-gene, *hlyA*-gene

Figure 1.4. Scheme of ISO procedure for the detection of *E. coli* O157 from food and feed samples (International Standards Organization, 2001a). *: incubation at its recommended time and temperature profile.

After isolation on selective agars, the presumptive colonies have to be confirmed using the production of indole as test parameter and serological *E. coli* O157 agglutination tests. The presence of the different virulence genes needs to be examined. This genetic characterisation has to be performed in a reference lab (International Standards Organization, 2001a).

1.2.3 Alternative detection methods

The reference methods used by many regulatory laboratories around the world are based on laborious, large volumes usage of liquid and solid media and reagents, and time consuming procedures both in operation and data collection. However, these methods remain as the "gold standards" in applied microbiology and are the preferable methods prescribed in the books of regulatory agencies nationally and internationally to the present time. The reference methods are still quite efficient and sensitive, cheap and specific and reliable. Last two decades, interest raised in the development of more "rapid" methods mainly based on molecular and immunological methods (Feng, 1996; Fung, 1994).

1.2.3.1 Modifications of conventional detection methods

By minor modifications to the reference (conventional) methods, the time to detection has reduced or the performance/convenience of the method has been improved. For example recently interest raised in the development of improved or one-step enrichment broths that allows enhanced resuscitation of the sub-lethally injured target cells and more rapid growth until detectable numbers enabling to reduce the total enrichment time. Various commercially available one step enrichment broths are available for *L. monocytogenes*: ONE Broth (Oxoid, Basingstoke, UK), *Listeria* special broth (Bio-Rad, Nazareth Eke, Belgium) and *Listeria* eXpress broth (BioMérieux, Marcy l'Etoile, France). In a similar manner, recently Bio-Rad launched the Rapid'*Salmonella* Short Protocol, were one single enrichment broth is used (BPW + *Salmonella* capsule) to shorten and to facilitate the enrichment step.

The recognition of colonies of presumptive target organisms has been facilitated by the introduction of chromogenic and fluorogenic media. These are microbiological growth media that contain enzyme substrates linked to a chromogen (colour reaction), fluorogen (fluorescent reaction) or a combination of both. The target population is characterized by enzyme systems that metabolize the substrate (e.g. sugar or amino acid) to release the chromogen/fluorogen. This results in a colour change in the medium and/or fluorescence under long wave UV light (e.g. with Wood's lamp = 365 nm). The incorporation of such fluorogenic or chromogenic enzyme substrates into a selective medium can eliminate the need for subculture and further biochemical tests to establish the identity of certain microorganisms (Greenwood *et al.*, 2005; Manafi, 1996; Manafi, 2000; Perry and Freydiere, 2007; Schonenbrucher *et al.*, 2008). Some chromogenic media have been acknowledged to improve significantly the performance of the isolation or enumeration of defined target organisms and they have therefore been taken up in the standardized ISO methods. For example the Agar *Listeria* according to Ottaviani and Agosti, abbreviated as ALOA, has been shown to provide typical colonies for *L. monocytogenes* and has been taken up in the ISO 11290-1/A1:2004 (International Standards Organization, 2004). Similar the Tryptone-bile-glucuronic chromogenic medium, abbreviated as TBX, medium has become the agar medium of choice for ISO 16649:2001 for the enumeration of β-glucuronidase positive *E. coli* (International Standards Organization, 2001b). A variety of chromogenic media are nowadays available from different manufacturers either for enumeration of coliforms or *E. coli*, or for isolation of pathogens such as *Salmonella*, *E. coli* O157, *Cronobacter sakazakii*, etc.

At the end of the detection method, various user-friendly identification kits have been introduced to replace the cumbersome classical confirmation step. BioMérieux developed a range of API® test kits for convenient identification of several groups of microorganisms. They are based upon galleries of miniaturized test tubes with dehydrated reagents which are easily inoculated with a cell suspension of the presumptive colony. The VITEK® cards (BioMérieux) are a variant on this and are part of a fully automated system that provides identification of organisms in a few hours. Other rapid identification kits, each in a different user-friendly format, are available; Crystal ID (Becton Dickinson, Erembodegem, Belgium), MicroLog 1, MicroLog 2, OmniLog® ID and MicroLog™ MicroStation™ (Biolog, Hayward CA, USA), the latter two also amenable to automation. For defined organisms, rapid confirmation kits were developed e.g. Oxoid Biochemical identification system-mono (Oxoid) for *L. monocytogenes* and *Listeria* spp.

1.2.3.2 Immunoassays

All immunoassays are based on the highly specific binding reaction between antibodies and antigens. The selection of an appropriate antibody (monoclonal or polyclonal) is the determinant factor for the method's performance. Usually, any positive result for pathogens obtained with immunoassays is considered as presumptive and requires further confirmation. Detection limit is approximately 10^4-

10^5 CFU/ml, depending upon the type of antibody and its affinity for the corresponding epitope meaning that for the enrichment step often a two-step procedure is needed. Several types of immunoassays are available in food diagnostics of which lateral flow devices (LFD), Enzyme Linked Immunosorbent Assays (ELISAs) & Enzyme linked fluorescent assays (ELFAs) are widely used. Immunomagnetic separation and concentration (IMS) assays have been developed as an aid in reducing the time for the enrichment step prior to detection, but will be discussed in the section regarding sample preparation.

Agglutination and latex agglutination are also based on the binding reaction of antibodies and antigens and are used to test an unknown organism against known antisera with aim of confirmation. These tests are able to indicate the presence of an analyte and can be used for serological confirmation. The term agglutination refers to the *in situ* produced macroscopic clumping. This reaction is generally rapid, does not dissociate and shows high affinity. Latex agglutination uses latex particles to amplify antigen-antibody agglutination reactions (Hazeleger *et al.*, 1992; Miller *et al.*, 2008; Niskanen *et al.*, 1991; van Griethuysen *et al.*, 2001).

- <u>*Lateral flow devices (LFD)*</u>

A LFD generally comprises a porous membrane, typically nitrocellulose, with an immobilized capture protein for the target analyte, forming a visible line in a viewing window after contact with the specified pathogen. For the test to be valid a control line should form in a second viewing window. In most devices, an antibody is commonly used as a capture protein, which specifically binds and captures a particular antigen if present in the sample. The test is fast (reading in terms of minutes) simple in use and in interpretation (Aldus *et al.*, 2003; Bird *et al.*, 1999). LFD such as e.g. Singlepath® can be used as a detection method, but also as a confirmation method.

- <u>*Enzyme Linked Immunosorbent Assay (ELISA) & Enzyme linked fluorescent assays (ELFA)*</u>

ELISA is a biochemical technique that couples an immunoassay with an enzyme assay. In most of the alternative methods a sandwich ELISA is used. The sandwich ELISA comprised different steps. Specific antibodies are affixed to the surface of the wells of a 96-well microtiter plate. The sample, with an unknown amount of target antigen, is added and allowed to bind to the affixed antibodies. Unbound antigen is removed by a washing step. In a second phase antibodies targeting the antigen are added again to the wells. This step is followed by the addition of an enzyme-labelled secondary antibody. This secondary antibody is allowed to bind to the previously added antibody. Washing steps are included to remove non-bound secondary antibodies. In the final step a substrate is added that the enzyme can convert to a detectable signal. The ELISA detection itself takes only 2-3 hours (Cantoni *et al.*, 1998; Croci *et al.*, 2001; Kumar *et al.*, 2008; Lilja and Hanninen, 2001; Scheu *et al.*, 1999; Szabo *et al.*, 2008; Thacker *et al.*, 1996).

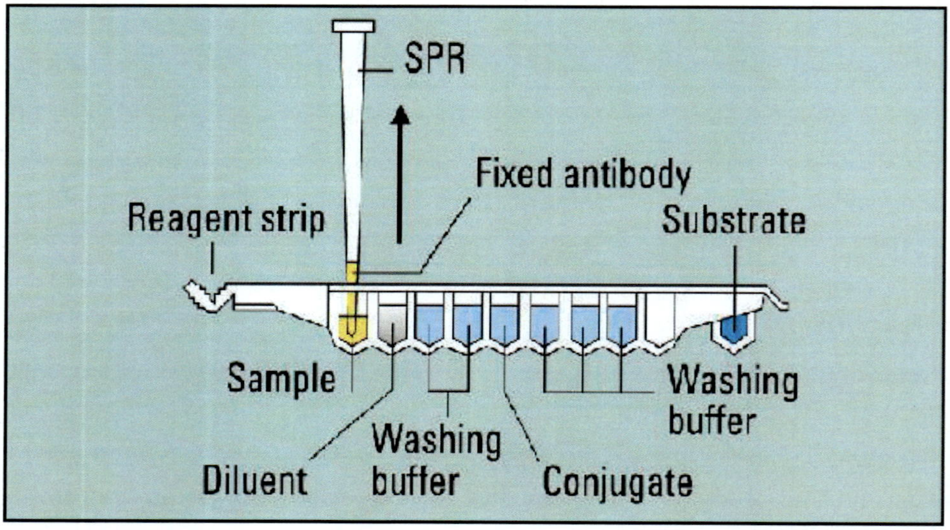

Figure 1.5. Schematic representation of the principle of VIDAS® an ELFA method (adopted from BioMérieux, 2001).

Nowadays many ELISA tests are set available as robotized automated systems to reduce the hands-on time but also to improve the reproducibility and standardization of each step of the assay. While many ELISA-based methods rely on chromogenic substrates for end-point detection of the target antigen, ELFA employs fluorescence for end-point detection. The VIDAS® system developed by BioMérieux is an entirely automated analysis using the dedicated VIDAS® equipment and corresponding test tips (holding the antibodies) and test strips for inserting the sample and holding the reagents (Figure 1.5.). A run of VIDAS® takes 45-120 min depending on the target pathogen (Cohen and Kerdahi, 1996; Keith, 1997; Sewell *et al.*, 2003).

1.2.3.3 <u>Bacteriophage based detection methods</u>

Bacteriophages are viruses infecting bacteria. Phages are extremely host-specific. Most bacteria can be infected by particular phages and it is common that a given phage can recognize and infect only one or a few strains or species of bacteria. The specificity of these phages is partly mediated by tail-associated proteins that distinctively recognize surface molecules of susceptible bacteria (Kretzer *et al.*, 2007).

- *<u>Alaska AK Phage™ detection method</u>*

The Alaska AK Phage™ method comprises IMS step for capturing the target bacteria from pre-enrichment broths and a bacteriophage step to specifically lyse viable bacteria and release the intracellular adenylate kinase (AK). Then, an amplification step based on AK-mediated conversion of ADP to ATP is done. Finally, the resulting ATP is detected by sensitive bioluminescence.

The standard "ATP bioluminescence" test (Bell *et al.*, 1996; Chen and Godwin, 2006; Fukuda *et al.*, 2005; Niza-Ribeiro *et al.*, 2000; Samkutty *et al.*, 2001; Siragusa *et al.*, 1995) is with this technique improved in sensitivity because of the amplification step by AK and because of the combination with the capture by IMS. The specificity of the test is provided by the use of the antibodies during the IMS capture as well as by the use of a specific bacteriophage to lyse the cells. Results of the test, as such are obtained within hours (Blasco *et al.*, 1998; Wu, 2008) although also a prior enrichment takes 6-24 h depending upon the type of microorganism.

- *<u>Bacterial bioluminescence</u>*

Many marine microorganisms are capable of emitting light due to the presence of so-called *lux*-genes. The introduction of these genes into highly specific bacteriophages by genetic engineering has resulted in the development of rapid and sensitive methodologies for detection of bacterial pathogens in food. Bacteriophages possessing *lux*-genes infect target bacteria, and this infection results in expression of the light-emitting genes. Target bacteria will luminescent upon the addition of an exogenous substrate. The time to obtain a result is usually less than 1 hour and the luminometers can detect as few as 500 bacteria. The disadvantage of the technique is that there are some luminescent non-microbial substances in the biological samples that can increase the intensity of the measured light (Baker *et al.*, 1992; Dostalek and Branyik, 2005; Goodridge and Griffiths, 2002; Meighen, 1991; Meighen, 1993; Stewart and Williams, 1992).

1.2.3.4 Microscopy based detection methods

Flow cytometry quantitatively measures the optical characteristics of cells (or other particles) as they are presented separately in front of a focused light beam (from a high-pressure mercury vapour lamp or an assortment of lasers). In order to present cells one by one they are introduced into a fast flowing fluid stream termed the sheath flow. As particles pass through the light beam three parameters are measured using photomultiplier tubes, the forward scatter, the side scatter and fluorescence. The amount of light scattered forward and at right angles by any particle tends to increase with cell size. In addition, the cell refractability is related to surface properties and internal structures and these affect forward and side scatters. Natural fluorescence (autofluorescence) is emitted by cellular components such as flavin nucleotides, pyridine and photosynthetic pigments. Using conventional flow cytometry (without labelling), most microorganisms are optically too similar to resolve from each other or from debris. To overcome this limitation, labelling techniques are used. Fluorescent dyes can be used to probe the viability and metabolic state of microorganisms.

For routine analyses of milk quality, an automated instrument was developed. This flow cytometry method uses ethidium bromide (intercalating with DNA) to stain bacteria in milk. The disturbing milk components are reduced and dispersed by treatment with detergent and enzymes at 50°C and provides a result after 8 min (Suhren and Reichmuth, 1997; Suhren and Walte, 1998).

Flow cytometry is often used in combination with live-dead staining. This staining is based on membrane integrity of cells. Most of the membrane integrity assays use nucleic acid stains. A wide diversity of nucleic acid stains is available but propidium iodide (PI) is most commonly used. In combination with SYTO-9, PI is able to distinguish between live and dead cells. SYTO-9 penetrates inside cells with both intact and damaged membranes, whereas PI only penetrates cells with damaged membranes (Joux and Lebaron, 2000).

An alternative is to use fluorescent stains such as ChemChrome (AESChemunex) that indicate metabolic activity. Viable cells are detected by their ability to cleave (by use of esterase activity) the nonfluorescent precursor fluorescein diacetate (FDA) to release a calcein or fluorescein (fluorescent) that is kept intra-cellular in intact cells (negative charge at neutral pH i.e. indicating an intact cell membrane) and thus give rise to a fluorescent signal (Berney *et al.*, 2008; Flint *et al.*, 2007; Tanaka *et al.*, 2000).

Any method that simply compares flow cytometry counts with plate counts (the 'golden' standard) should be aware that deviation between flow cytometry and plate counts may occur due to the presence of viable but non cultivable cells. In general flow cytometry can be described as an automated microscopy that has the advantages of automation, objectivity and speed as flow cytometry can analyse many thousands of cells per second but the detection limit is rather high (approximately 10^4 CFU/ml) and interference by the food matrix (suspended particles and autofluorescence) may be encountered (Flint *et al.*, 2006; Gunasekera *et al.*, 2003).

1.2.3.5 <u>Molecular detection methods</u>

The ability of two single-stranded DNA or RNA molecules *in vitro* to hybridize by specific base pairing is the basis of all molecular based detection methods. The similarity of the two sequences determines their degree of hybridisation. Two methods, i.e. FISH and PCR are in particular relevant for detection of bacteria in food. The selection of a specific DNA sequence, to serve as a probe or primer, along with the conditions for hybridisation, is the determinant factor for the specificity of these molecular methods.

- *Fluorescent* in situ *hybridisation*

Fluorescence *in situ* hybridisation (FISH) with ribosomal RNA (rRNA) targeted oligonucleotide probes is the most commonly applied technique among the 'non-PCR-based' molecular techniques. The choice to target RNA instead of DNA results in a more sensitive technique (higher copy numbers available) and the link to viability.

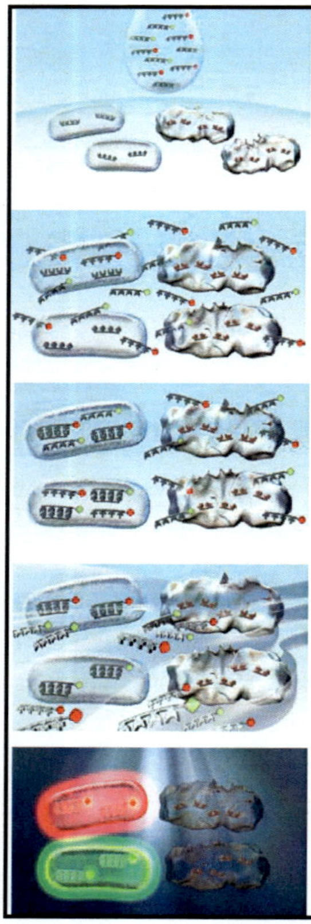

1. Contact of gene probes with the samples.

2. Gene probes diffuse into all cells.

3. Gene probes bind to their target sequences within the cells. No binding in dead cells.

4. Unbound gene probes are removed by washing step.

5. Labelled viable cells start to illuminate and can be identified and quantified.

Figure 1.6. Schematic representation of the principle of Vermicon Identification Technology (VIT) (adapted from http://www.vermicon.de, 10/09/2009).

In the elaboration of FISH, microbial cells are treated with appropriate chemical fixatives and then hybridized under stringent conditions on a glass slide or in solution with oligonucleotide probes. Generally, these probes are 15-25 nucleotides in length and are labelled covalently at the 5' end with a fluorescent dye. After stringent washing, to remove unbound probe, specifically stained cells are detected via epifluorescence microscopy (Figure 1.6.). As only a limited number of microscopic fields (correlating to a small volume) are screened for fluorescence resulting in a limit of detection of approximately 10^4 CFU/ml. After prior enrichment (usually overnight) to attain these levels of detection, results are available in three hours while the hands-on time required per analyze is only a few minutes (Bottari *et al.*, 2006; Moter and Gobel, 2000; Schmid *et al.*, 2005; Stephan *et al.*, 2002; Stephan *et al.*, 2003; Wagner *et al.*, 2003).

- *Conventional and real-time Polymerase Chain Reaction*

The PCR technique, first described by Kary Mullis in the mid-1980s is a three-step cyclic *in vitro* procedure based on the ability of the DNA polymerase to copy a strand of DNA (Mullis and Faloona, 1987). The start and the end of the region of interest are specified by the choice of primers. Primers are short oligonucleotides, usually 20-30 nucleotides in length, whose sequence matches the end of the region of interest. Amplification takes place over a number of cycles. During each cycle the double-stranded DNA template is denaturated by heating to produce single-strands. The reaction mixture is then cooled allowing the primers to bind to the single-strands. This provides an active site for thermo-stable DNA polymerase which synthesises the complementary strand, producing again double-stranded DNA. In subsequent cycles primers will bind to both the original DNA and the newly synthesised DNA resulting in an exponential increase in the numbers of copies. The presence of even 1 copy of the template within the reaction mixture can be detected within a couple of hours as about a million-fold of copies are created (Hanna *et al.*, 2005). Usually the region amplified is shorter than 3 kb although larger fragments can be amplified with less efficiency (Ercolini, 2004; Harris and Griffiths, 1992; Malorny *et al.*, 2003b; Settanni and Corsetti, 2007).

The results of PCR are traditionally (conventional PCR) detected by agarose gel electrophoresis and staining. This enables the amplified DNA to be visualised as bands differing in size. The size of the bands can be determined by comparison with standard molecular weight markers. Presence of a band(s) of expected size indicates a positive result. Specificity of the bands may be further identified by sequencing.

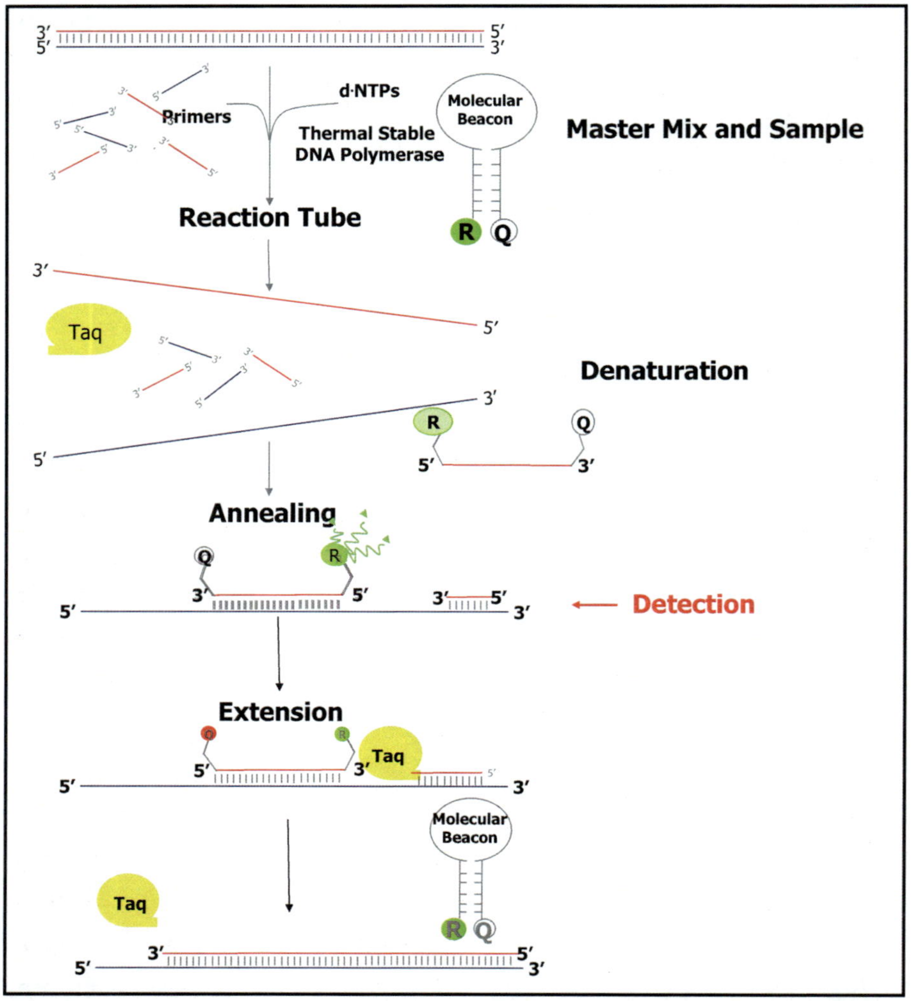

Figure 1.7. Schematic representation of PCR principle based on molecular beacons (adapted from Bio-Rad).

In the early 1990's, the "second" generation of PCR technologies was introduced by the use of fluorescent double stranded DNA dyes e.g. SYBR® Green or DNA probes e.g. Molecular beacons® (Figure 1.7) or TaqMan® probes and a fairly new type of chemistry is the scorpion probe (Hanna *et al.*, 2005). Real-time PCR enables both the detection and quantification of a nucleotide signal by continuously measuring a fluorescent reporter during the PCR reaction. Real-time PCR enables the on-line detection and might be used for quantification as well. Not only the fact that PCR is no longer a black box reaction, but also the fact that real-time PCR consists of a closed-tube format, reducing the risk of contamination potentially leading to false-positive results, boosted the potential application of PCR also in routine labs.

The terminology "rapid" and "sensitive" in the description of the PCR technique as a detection method for food-borne pathogens is relative. PCR as such is taking only approximately 30-90 minutes. However, PCR methods for detection of pathogens in foods involve a 6-8 h up to 24 h prior enrichment step before execution of PCR as only a small volume (0.1-1 ml) is processed to extract DNA. Shortened enrichment procedures, overlooking the need to repair stressed and/or sub-lethal injured cells may result in a true risk of not reaching the detection limit of the subsequent PCR technique leading to false negative results. If a further reduction of time to detection is needed, further progress on elaboration of sample preparations procedures should be made. Focus is actually put on concentration protocols of bacteria or DNA from larger volumes (250 ml) e.g. the use of bacteriophage-based capture (Hagens and Loessner, 2007), devices such as Pathatrix® (Matrix MicroScience Inc.) enabling up-scaling of present available capture or extraction methods (Rotariu *et al.*, 2005; Wu *et al.*, 2004), etc.

Nevertheless, during concentration of cells (or DNA), care has to be taken not to concentrate potential interfering factors from the food matrix as well. Since this will jeopardize the efficient functioning of the PCR reaction and in turn might lead to false negative results. The inclusion of an internal control is recommended to reveal inhibition of the PCR reaction (Malorny *et al.*, 2003b; Rodriguez-Lazaro *et al.*, 2007; Rodriguez-Lazaro and Hernandez, 2006).

1.2.3.6 Future trends

- *Further developments of Multiplex PCR*

To rapidly detect multiple microorganisms in a single reaction, simultaneous amplification of more than one locus is required. This methodology is referred to as multiplex PCR (MPCR) in which several specific primer sets are combined into a single PCR assay. MPCR is undoubtedly useful to rapidly identify several isolates. However, the MPCR technique is particularly versatile and it has been applied successfully in many fields of microbiology for faster identification, direct detection and differentiation of microorganisms with and/or without previous isolation and cultivation. Recently MPCR detection kits are also commercially available e.g. GeneDisc EHEC (Beutin *et al.*, 2009; Kawasaki *et al.*, 2009; Settanni and Corsetti, 2007). The technique has a lot of potential however the limitations regarding the amount of different targets in the multiplex should be recognised.

- *Microarrays*

Microarrays or gene chips provide a miniaturized system for the simultaneous analysis of hybridization of fluorescent-labelled single-strand nucleotide chains to an array of oligonucleotide probes immobilized on a support such as glass or a synthetic membrane. PCR amplification is often used prior to hybridization to increase sensitivity of detection (Keramas *et al.*, 2003). DNA microarrays may be very useful for detecting multiple bacteria simultaneously on a single glass slide (Hong *et al.*, 2004; Ikeda *et al.*, 2006; Kostrzynska and Bachand, 2006; Zhou, 2003). The complexity of the food matrix is a major drawback of microarrays to be used as a detection method. Therefore microarrays are in the frame of food diagnostics rather described, as a tool for identification, genotyping and pathotyping (detection of appropriate virulence factors) or characterisation of bacterial isolates. Selected genes often cover functional, virulence or stress genes. The most common application of microarray technology is differential gene expression analysis with the main objective being comparison of the expression levels of selected genes under control and test (often stress) conditions. Since DNA arrays allow simultaneous measurements of thousands of interactions between mRNA-derived target molecules and genome-derived probes, they are rapidly producing enormous amounts of raw data never before encountered by biologists. The analysis of data on this scale is a major current challenge and therefore needs permanent attention to the issue of sample preparation and genetic material purification for successful and reliable analysis.

However, also the technique of microarrays evolves and becomes more accessible for application in routine food diagnostics. Recently a microarray system, the

Premi®System *Salmonella* system has been brought to the market by DSM Nutritional Products (Heerlen, The Netherlands) aiming at identification of the most common serovars of *S. enterica*. Starting from a presumptive *Salmonella* colony, results on the serotype/genotype of *Salmonella* are obtained within 8 h. The system allows single-tube processing of the samples and requires no peculiar technical skill. As such it shows interesting potential to be used for routine laboratory testing (Wattiau *et al.*, 2008).

- *Biosensors*

Biosensors are defined as analytical devices that combine bio-specific recognition systems with physical or electrochemical signalling. Biosensors for the detection of pathogens in the food industry consist of immobilized biologically active material, like enzymes, antibodies, antigens or nucleic acids, in close proximity to a receiving transducer unit. Target recognition results in the generation of an electrical, optical or thermal signal that is proportional to the concentration of target molecules. Examples of physical signals, which can report the presence of molecules, are fluorescence signals from dyes, electric fields from molecular charges, or mass changes or refractive index changes from the adsorption of molecules onto sensor surfaces (Maraldo and Mutharasan, 2007; Nordstrom *et al.*, 2008).

Biosensors have the potential to shorten the time between sampling and results, but their future lies in reaching selectivity and sensitivities comparable to established methods at a fraction of the cost. The use of biosensors allows both miniaturization and automation. It allows working with sample volumes in the range of nanolitres or less, which implies that, the cost of reagents will not be high. Also the performance of multi-analyte analysis is enabled in the same device, which again shortens analysis time (Ivnitski *et al.*, 1999; Lazcka *et al.*, 2007; Rasooly and Herold, 2006; Ricci *et al.*, 2007; Skottrup *et al.*, 2008; Zhang *et al.*, 2009).

Piezoelectric-excited millimetre-sized cantilever (PEMC) sensors described by Campbell *et al.* in different papers (Campbell *et al.*, 2007c; Campbell *et al.*, 2007a; Campbell *et al.*, 2007b; Campbell and Mutharasan, 2005; Campbell and Mutharasan, 2008) consist of a piezoelectric and a borosilicate glass layer with a sensing area of 4 mm^2. An antibody specific to *E. coli* O157:H7, *Bacillus anthracis*, *Cryptosporium parvum* or *Staphylococcus* enterotoxin B was immobilized on PEMC sensors and allowed the real-time detection of these pathogens/toxins. In the specific case of *E. coli* O157:H7, Campbell *et al.* (2007c) showed that the PEMC sensor could detect the pathogen in beef samples under flow conditions in real time without sample preparation. The sensor was able to detect about 50–100 cells/ml of meat particle containing broths.

1.2.4 Validation of alternative detection methods

It is of the utmost importance in the framework of verification and validation of food safety management systems, as well as regarding governmental monitoring systems, that the results obtained with these "rapid" alternative tests are reliable and that all parties involved agree with and accept the methodology employed (Jacxsens *et al.*, 2009). Food business operators have the possibility to use analytical methods other than the reference methods, in particular more rapid methods, as long as the use of these alternative methods provides equivalent results. According to the EU Regulation 2073/2005 (EU 2005), the use of alternative analytical methods is acceptable when the methods are validated in accordance with the protocol set out in ISO standard 16140:2003 for method validation (International Standards Organization, 2003b). The use of validated methods is also a prerequisite for (service) laboratories accredited according to ISO 17025:2005 (International Standards Organization, 2005a) or food companies whose labs are an integrated part of the companies' food safety management systems and are subjected to third parties audits that require quality assurance-based test results.

There are several validation bodies namely, AOAC (Association of Official Analytical Chemist, USA), AFNOR (Association Français de Normalisation, France), NordVal (part of the Nordic Committee on Food Analysis, Norway) and MicroVal (European Validation and Certification Organisation, Europe) that issue validation certificates for alternative methods. ISO 16140:2003 is a technical protocol for the validation of alternative methods in the field of microbial analysis of food, animal feeding stuff and environmental and veterinary samples (Betts and Rentenaar, 1998; Rentenaar, 1996). Although some differences exist in the different validation protocols, the rules for the validation process of AFNOR, MicroVal as well as NordVal are based upon ISO 16140:2003. AOAC validation studies also comply to a large extent with the demands of ISO 16140:2003, but divides up certification into either, AOAC Performance Tested Methods (which do not include an inter-laboratory trial) and AOAC Official Method (which do include an inter-laboratory trial). In addition, whereas AFNOR, NordVal and MicroVal usually certify a method for all foods if at least five categories of foods were taken up in the validation study, AOAC rather mentions validation for a summed up variety of foods taken up in the validation study. It should be clear that, notwithstanding the external validation of a method, when taking up a method in the own lab, it is recommended that this method should be further subjected to a secondary validation (also referred to as verification) to demonstrate that an established method functions in the user's hands with the food types analysed within one's own lab.

In its widest sense, validation is showing the fitness of purpose of a method i.e. the degree to which data produced by a method enables a user to make technically and administratively correct decisions for a stated purpose (e.g. performance measurement of implemented food safety management systems). Usually, method validation is defined as the establishment of the performance characteristics of a method and provision of objective evidence for this. For qualitative methods (presence/absence testing) the accuracy, specificity, sensitivity and detection limit are the major technical characteristics to be checked. For quantitative methods (enumeration) also the quantification limit and the linearity should be checked.

In the frame of ISO 16140:2003, a validation encompasses a method's comparison study especially looking for the degree of correspondence between the response obtained by the reference method and the response obtained by the alternative method on identical samples. An inter-laboratory study of the method in question is also part of the full validation study. Interlaboratory testing further provides an indication of the precision and the robustness of a method. Note that in validation of an alternative method it is essential to understand the basis of new methods, i.e. to have an unambiguous description/definition of the target organism and the (principle of the) method. So that if differences are found when comparing a new method, often based upon a different principle of detection, with an established one, often based on cultivability in the presence of selective compounds and phenotypic characteristics, explanations for these differences can be sought.

Commercially available rapid detection methods such as common used ELISA and PCR kits have reduced the time and the labour involved in the analysis of food products, but they still rely on the enrichment step. It can't be neglected that the rapid detection method significantly reduce the overall assay time compared to the conventional methods and in the scope of negative screening these rapid method have for sure their advantage in time (Bohaychuk *et al.*, 2005).

Table 1.3 provides an overview of "rapid" alternative detection methods with at least one validation certificate at the time of writing. The time to detection is highlighted as it is a major factor in the selection of a method by both food industry and competent authorities. MicroVal is not incorporated in Table 1.3. At the moment of writing no alternative detection method of regarding "the big 4" was validated yet.

From Table 1.3 it is clear that only three alternative detection methods for *Campylobacter* spp. are validated. Singlepath® *Campylobacter* and BAX® system for detection of *Campylobacter* spp. are AOAC-*PTM* validated and allow detection after two days of enrichment. The Landmännen Danpo PCR method for *Campylobacter* spp. is validated by NordVal and allows detection after one day of enrichment, resulting in a three days reduction compared to ISO method.

For the alternative detection methods of *L. monocytogenes*, based on chromogenic media, four out of seven allow detection within two days, compared towards the four days of ISO method. The three other chromogenic media are developed with the aim to serve as second media in the ISO method. For detection of *Salmonella* spp. two out of three chromogenic media decrease the time to obtain a result with one day, while the chromogenic media developed for detection of *E. coli* O157 do not decrease the time to detection (compared to ISO method), but facilitate the reading and the detection of the pathogen and the workflow (one isolation plate and one confirmation method). Also TRANSIA® PLATE *L. monocytogenes*, a LFD, allows detection within two days, similar for VIDAS® LMO method.

Most of the LFD developed for *Salmonella* spp. allow detection within 2 days (4/5) and the LFD for *E. coli* O157 within 24 h. The number of detection methods for *Salmonella* which are able to give results within 24 h is limited as seen by (Eijkelkamp *et al.*, 2009) and even more limited if validation is required as can be noticed from Table 1.3. VIDAS® up is an improvement regarding time to detection compared to the standard VIDAS® for *E. coli* O157. Table 1.3 depicts five validated real-time PCR detection kits for *E. coli* O157, all of them relying on an enrichment step of maximum 24 h although shorter time for enrichment are possible depending on the food sample.

Table 1.3 shows that for the chromogenic media most of the validation studies were performed by AFNOR, in particular for *L. monocytogenes*. Regarding immunological detection method, the LFD are only validated by AOAC and not by the European validation bodies. This may reflect different preferences for testing methods in the European Union versus the U.S. This can be further noticed from the fact that most of the ELISA methods detect *Listeria* spp. instead of *L. monocytogenes*, with the exception of TRANSIA® PLATE *L. monocytogenes* which is the only validated ELISA method that can detect *L. monocytogenes*. In the U.S. the attitude is to test for *Listeria* spp. (zero-tolerance) while in Europe there is a certain tolerance regarding *L. monocytogenes* in certain foods. The VIDAS® (ELFA) can also distinguish between *L. monocytogenes* and other *Listeria* spp. In most of the cases the ELISA methods for *Salmonella* spp. are validated by AFNOR and AOAC, but nowadays also NordVal provided certificates.

Technical performance characteristics such as specificity, sensitivity and detection limit are not taken up in Table 1.3 as they are inherently connected to the fact that the methods are certified. Certified methods are being found to be equivalent to the reference method i.e. detection limit of approximately 5 CFU/g. All methods involved in Table 1.3 are proprietary rapid methods (except the Lantmännen Danpo A/S PCR methods), often amenable to automation and with a user-friendly interface, ready to be used by lab technicians with a basic training in microbiology, with good notions of aseptical-working procedures and culturing and handling of microorganisms in the lab. When manipulating pathogens extra training concerning bio-safety issues is needed.

Table 1.3. Overview of alternative detection methods validated by at least on the official validation bodies.

		L. monocytogenes					*Salmonella* spp.					*E. coli* O157					*Campylobacter* spp.					Remarks
Classical cultural method (ISO method)		PE: 24 ± 2 h + E : 48 ± 2 h + D : 24 ± 2 h (Total: ± 4 days)					PE: 18 h + E : 24 ± 3 h + D : 24 ± 3 h (Total: ± 3 days)					PE: 6 up to 18-24 h + IMS + D : 18-24 h (Total: ± 2 days)					PE: 4-6 h + E : 44 ± 4 h + D : 44 ± 4 h (Total: ± 4 days)					Identification and confirmation not included
Name	Website	Time (h)	AFNOR	NordVal	MicroVal	AOAC*	Time (h)	AFNOR	NordVal	MicroVal	AOAC*	Time (h)	AFNOR	NordVal	MicroVal	AOAC*	Time (h)	AFNOR	NordVal	MicroVal	AOAC*	
CHROMOGENIC AND OTHER ALTERNATIVE MEDIA																						
chromID™	www.biomerieux.com	**	X				***					***					***					Multi-purpose agar
RAPID'	www.bio-rad.com	PE: 24 ± 2 + D : 24 ± 2	X	X		x_2	PE: 18 ± 2 + E : 6 -26 + D : 24 ± 2	X	X		x_2	PE: 16 - 24 + IMS D : 24 ± 2	X			x_2	/					Multi-purpose agar
							PE: 18 ± 2 + D : 24 ± 2	X	X													Buffered Peptone + *Salmonella* Capsule
AL Agar	www.bio-rad.com	PE: 24 ± 2 + D : 24 ± 2	X				NA					NA					NA					
Precis™ (Inhibigen™ technology)	www.oxoid.com www.remel.com	E : 22 - 26 + D : 22 - 26	X				E : 18 + D : 24	X			x_2	/					/					ONE broth + Brilliance™ agar Multi-purpose agar
ALOA® One Day	www.aeschemunex.com	PE: 24 ± 2 + D : 24	X				NA					NA					NA					
COMPASS® Agar	www.solabia.com (Biokar Diagnostics)	PE: 24 ± 2 + D : 24 ± 2 or **	X				***					/					/					Multi-purpose agar *** confirmation in SESAME *Salmonella* test
CHROM agar™	www.chromagar.com	**	X			x_2	**				x_2	**				x_2	/					Multi-purpose agar

		L. monocytogenes					*Salmonella* spp.					*E. coli* O157					*Campylobacter* spp.					Remarks
Classical cultural method (ISO method)		PE: 24 ± 2 h + E : 48 ± 2 h + D : 24 ± 2 h (Total: ± 4 days)					PE: 18 h + E : 24 ± 3 h + D : 24 ± 3 h (Total: ± 3 days)					PE: 6 up to 18-24 h + IMS + D : 18-24 h (Total: ± 2 days)					PE: 4-6 h + E : 44 ± 4 h + D : 44 ± 4 h (Total: ± 4 days)					Identification and confirmation not included
Name	Website	Time (h)	AFNOR	NordVal	MicroVal	AOAC*	Time (h)	AFNOR	NordVal	MicroVal	AOAC*	Time (h)	AFNOR	NordVal	MicroVal	AOAC*	Time (h)	AFNOR	NordVal	MicroVal	AOAC*	
SMS®	www.aeschemunex.com	NA					PE : 18 ± 2 + D : 24 ± 2	X				NA					NA					Semi-solid medium
SESAME *Salmonella* TEST	www.solabia.com (Biokar Diagnostics)	NA					PE: 18 ± 2 + D: 24 ± 3	X				NA					NA					
IMMUNOASSAYS																						
1. LFD																						
Singlepath®	www.emdchemicals.com (microbiology)	***					PE: 18 - 24 + E : 24 + D : 0.5				x_2	18-24 PE + 0.5 D				x_2	PE: 4 + E : 44 + D : 0.5				x_2	Only for *C. jejuni* and *C. coli*
DuPont™ Lateral Flow Systems	www.qualicon.com	*Listeria* spp.					PE: 5 - 24 + E : 18 - 24 + D : 0.17				x_2	PE: 8 - 18 + D : 0.17				x_2	/					
RapidChek® SELECT™	www.sdix.com	*Listeria* spp.					PE: 18 ± 2 + E : 18 ± 2 + D : 0.17				x_2	PE: 8 - 18 + D : 0.17				x_2	/					For *Salmonella* sliced cooked turkey E : 7 ± 1
VIP (Gold)	www.biocontrolsys.com	*Listeria* spp.					PE: 6 - 26 + E : 16 - 24 + EE: 5 - 8 + D : 0.17				x_1	EHEC					/					

		L. monocytogenes					***Salmonella* spp.**					***E. coli* O157**					***Campylobacter* spp.**					**Remarks**
Classical cultural method (ISO method)		PE: 24 ± 2 h + E : 48 ± 2 h + D : 24 ± 2 h (Total: ± 4 days)					PE: 18 h + E : 24 ± 3 h + D : 24 ± 3 h (Total: ± 3 days)					PE: 6 up to 18-24 h + IMS + D : 18-24 h (Total: ± 2 days)					PE: 4-6 h + E : 44 ± 4 h + D : 44 ± 4 h (Total: ± 4 days)					Identification and confirmation not included
Name	Website	Time (h)	AFNOR	NordVal	MicroVal	AOAC*	Time (h)	AFNOR	NordVal	MicroVal	AOAC*	Time (h)	AFNOR	NordVal	MicroVal	AOAC*	Time (h)	AFNOR	NordVal	MicroVal	AOAC*	
TRANSIA® CARD	www.biocontrolsys.com	*Listeria* spp.					***					E : 20 - 24 + D : 0.12				x_2	/					
Reveal®	www.neogen.com	*Listeria* spp.					PE: 2 - 4 + E : 18 + D : 0.25				x_2	E : 20 + D : 0.25				x_1	/					E : 8 for raw meat and vegetables
FoodChek™	www.vaccitest.com	/					/					E : 6 + D : 1				x_2	/					Based on nanotechnology (magnetic beads)
2. ELISA																						
RAPIDYME ELISA	www.bio-art.be	*Listeria* spp.					PE : 18 ± 2 + E : 24 ± 3 + D : 1.5	X				/					/					
LOCATE®	www.r-biopharm.com	*Listeria* spp.					E : 48 + D : 1.5				x_1	/					/					
Assurance EIA	www.biocontrolsys.com	*Listeria* spp.					PE: 24 ± 2 + E : 5 - 24 + EE: 5 - 18 + D : 2.5				x_1	EHEC					/					Instrumental read

		L. monocytogenes					*Salmonella* spp.					*E. coli* O157					*Campylobacter* spp.					Remarks
Classical cultural method (ISO method)		PE: 24 ± 2 h + E : 48 ± 2 h + D : 24 ± 2 h (Total: ± 4 days)					PE: 18 h + E : 24 ± 3 h + D : 24 ± 3 h (Total: ± 3 days)					PE: 6 up to 18-24 h + IMS + D : 18-24 h (Total: ± 2 days)					PE: 4-6 h + E : 44 ± 4 h + D : 44 ± 4 h (Total: ± 4 days)					Identification and confirmation not included
Name	Website	Time (h)	AFNOR	NordVal	MicroVal	AOAC*	Time (h)	AFNOR	NordVal	MicroVal	AOAC*	Time (h)	AFNOR	NordVal	MicroVal	AOAC*	Time (h)	AFNOR	NordVal	MicroVal	AOAC*	
Assurance Gold EIA	www.biocontrolsys.com	/					PE: 6 - 26 + E : 6 - 20 + EE: 6 - 20 + D : 2				x_1	/					***					Visual or instrumental read possible
RayAL Selecta	www.rayal.com	*Listeria* spp.					PE: 6 - 10 + E : 18 - 24 + D : 2	X	X			/					/					Automation of reading possible
RayAL Optima	www.rayal.com	*Listeria* spp.					PE: 16 - 20 + E : 18 - 24 + D : 2	X	X		x_2	/					/					Automation of reading possible
3M™ Tecra™ VIA	www.3m.com	*Listeria* spp.					PE: 18 - 22 + E : 16 - 20 + EE: 6 - 8 + D : 2.5				x_1	E : 18 - 24 + D : 2				x_2	***					
3M™ Tecra™ Ultima™	www.3m.com	/					PE: 16 - 20 + E : 10 - 24 + D : 2.5	X			x_1	/					/					

		L. monocytogenes					***Salmonella* spp.**					***E. coli* O157**					***Campylobacter* spp.**					**Remarks**
Classical cultural method (ISO method)		PE: 24 ± 2 h + E : 48 ± 2 h + D : 24 ± 2 h (Total: ± 4 days)					PE: 18 h + E : 24 ± 3 h + D : 24 ± 3 h (Total: ± 3 days)					PE: 6 up to 18-24 h + IMS + D : 18-24 h (Total: ± 2 days)					PE: 4-6 h + E : 44 ± 4 h + D : 44 ± 4 h (Total: ± 4 days)					Identification and confirmation not included
Name	Website	Time (h)	AFNOR	NordVal	MicroVal	AOAC*	Time (h)	AFNOR	NordVal	MicroVal	AOAC*	Time (h)	AFNOR	NordVal	MicroVal	AOAC*	Time (h)	AFNOR	NordVal	MicroVal	AOAC*	
3M™ Tecra™ Unique	www.3m.com	*Listeria* spp.					PE: 16 - 20 + D : 7	X			x_1	/					***					Automation possible
TRANSIA® PLATE (Gold)	www.biocontrolsys.com	PE: 20 - 26 + E : 20 - 48 + D : 2.5	X				PE: 18 ± 2 + E : 21 ± 3 + D : 2.5	X	X		x_2	***					***					Automation possible. with TAG 24 E: 4 - 5
RIDASCREEN®	www.r-biopharm.com	/					PE: Overnight + E + D : 1.25	X		X		/					/					
3. ELFA																						
VIDAS®	www.biomerieux.com	PE: 24 - 26 + E: 24 - 26 + D : 1.17	X			x_2 x_1	PE: 16 - 20 + E: 22 - 26 + D : 1	X			x_2 x_1	PE: 6 - 7 + E : 18 + D : 1	X			x_2	***					Shorter validated protocols for limited foods (see website)
4. Others																						
1-2 test	www.biocontrolsys.com	NA					E : 24 ± 2 + D : 14				x_1	NA					NA					Motile *Salmonella* spp.
Oxoid *Salmonella* Rapid test	www.oxoid.com	NA					E : 18 + D : 24	X			x_2	NA					NA					

		L. monocytogenes					*Salmonella* spp.					*E. coli* O157					*Campylobacter* spp.					Remarks
Classical cultural method (ISO method)		PE: 24 ± 2 h + E : 48 ± 2 h + D : 24 ± 2 h (Total: ± 4 days)					PE: 18 h + E : 24 ± 3 h + D : 24 ± 3 h (Total: ± 3 days)					PE: 6 up to 18-24 h + IMS + D : 18-24 h (Total: ± 2 days)					PE: 4-6 h + E : 44 ± 4 h + D : 44 ± 4 h (Total: ± 4 days)					Identification and confirmation not included
Name	Website	Time (h)	AFNOR	NordVal	MicroVal	AOAC*	Time (h)	AFNOR	NordVal	MicroVal	AOAC*	Time (h)	AFNOR	NordVal	MicroVal	AOAC*	Time (h)	AFNOR	NordVal	MicroVal	AOAC*	
BACTERIOPHAGE BASED METHODS																						
VIDAS-UP®	www.biomerieux.com	/					/					E : 6 - 24 + D : 1	X				/					
Alaska *fastr*AK™	www.alaskafooddiagnostics.com	*Listeria* spp.					PE: 16 + E : 2 + D : 0.08				x_2	***					/					
MOLECULAR BASED METHODS																						
1. Hybridisation																						
LUMIprobe 24	www.europrobe.com	E : 22 + D : 2	X				PE: 6 + E : 16 + D : 2	X				/					/					Automation for reading
Gene-Quence (=Gene-Trak®)	www.neogen.com	E : 40 - 48 + D : 2				x_2	E : 40 - 48 + D : 1.5				x_2 x_1	/					/					Based on rRNA Full automation possible
2. PCR																						
iQ-Check™	www.bio-rad.com	PE : 22 - 26 + D : 4	X	X		x_2	E : 16 - 22 + D : 4	X	□		x_2	E : 8 - 24 + D : 4	X			x_2	***					For *Salmonella* shorter protocols for meat (E = 8 h). □= Renewal of validation.

		L. monocytogenes					*Salmonella* spp.					*E. coli* O157					*Campylobacter* spp.					Remarks
Classical cultural method (ISO method)		PE: 24 ± 2 h + E : 48 ± 2 h + D : 24 ± 2 h (Total: ± 4 days)					PE: 18 h + E : 24 ± 3 h + D : 24 ± 3 h (Total: ± 3 days)					PE: 6 up to 18-24 h + IMS + D : 18-24 h (Total: ± 2 days)					PE: 4-6 h + E : 44 ± 4 h + D : 44 ± 4 h (Total: ± 4 days)					Identification and confirmation not included
Name	Website	Time (h)	AFNOR	NordVal	MicroVal	AOAC*	Time (h)	AFNOR	NordVal	MicroVal	AOAC*	Time (h)	AFNOR	NordVal	MicroVal	AOAC*	Time (h)	AFNOR	NordVal	MicroVal	AOAC*	
LightCycler®	www.roche.com	PE: 24 + (E : 24) + D : 2		X		x_2	PE: 16 - 20 + D : 2		X		x_2	E : 6 or 18 - 24 + D : 1.5		X		x_2	***					E for *L. monocytogenes* is optional
BAX® system	www.qualicon.com www.oxoid.com	E : 24 - 28 + D : 5	X			x_2 x_1	E : 16 - 24 + D : 4.5	X	X		x_2 x_1	E : 14 - 24 + D : 4.5	X			x_2	E : 24 - 48 + D : 2		X		x_2	48 E for *L. monocytogenes* = AOAC 8 E for *E. coli* O157 + IMS and 3.5 D = AOAC NordVal validated for cloacae swaps
GeneDisc	www.genesystems.fr	PE: 18 - 22 + D : 1.5	X				E : 8 - 12 or 16 - 20 + D : 1.5	X				E : 8 - 12 or 16 - 20 + D : 1.5					/					Also detection of non O157. Combination *E. coli* O157 - *Salmonella* spp. possible
ADIAFOOD®	www.aeschemunex.com	PE: 18 ± 2 + D : 2.5	○			x_2	PE: 18 ± 2 + D : 2.5	○			x_2	PE: 18 ± 2 + D : 2.5	○			x_2	/					(Warnex™) PE: 8 for *Salmonella* spp. and *E. coli* O157 in raw meat AOAC *PTM* for quantification of *Campylobacter* spp. in poultry rinses

		L. monocytogenes					***Salmonella* spp.**					***E. coli* O157**					***Campylobacter* spp.**					**Remarks**
Classical cultural method (ISO method)		PE: 24 ± 2 h + E : 48 ± 2 h + D : 24 ± 2 h (Total: ± 4 days)					PE: 18 h + E : 24 ± 3 h + D : 24 ± 3 h (Total: ± 3 days)					PE: 6 up to 18-24 h + IMS + D : 18-24 h (Total: ± 2 days)					PE: 4-6 h + E : 44 ± 4 h + D : 44 ± 4 h (Total: ± 4 days)					Identification and confirmation not included
Name	Website	Time (h)	AFNOR	NordVal	MicroVal	AOAC*	Time (h)	AFNOR	NordVal	MicroVal	AOAC*	Time (h)	AFNOR	NordVal	MicroVal	AOAC*	Time (h)	AFNOR	NordVal	MicroVal	AOAC*	
TaqMan® Detection kit	www.appliedbiosystems.com	***					PE: 1 + D : 4	X			x2	***					***					For *S. enterica*
R.A.P.I.D.® LT real-time PCR system	www.idahotech.com	*Listeria* spp.					PE: 16 + D : 0.5				x2	/					/					
Lantmännen Danpo A/S	www.lantmannen.com	/					ND		X			/					24 E + 3 D		X			(Krause *et al.*, 2006)
Assurance GDS™	www.biocontrolsys.com	PE: 30 + D : 2				x2	E : 18 - 24 + D : 2	X			x2	PE: 6.5 - 24 + D : 1.5				x1	/					IMS coupled with PCR detection

X: validation certificate
***: AOAC stand for AOAC-RI *Performance tested method* (PTM) and AOAC *Official Method of Analysis* (OMA)**
x_1 : AOAC-*OMA* validation certificate
x_2 : AOAC-*PTM* validation certificate
****: If detection protocol of alternative method is similar to ISO protocol**
*****: method exists for MO but is not validated**
/: does not exist
○: validation is pending
NA: not applicable: as example no ALAO agar for *Campylobacter* spp.
PE: pre-enrichment
E: enrichment
EE: extra enrichment (= post-enrichment)
D: detection
ND: not determined

1.3 Bottlenecks in rapid methods

Recovery of food-borne pathogenic bacteria from food and feed or from the processing environment has traditionally been a challenge. Any rapid detection system, while potentially very sensitive, must confront a number of obstacles inherent with sampling complex food and feed matrices. Relative low levels of pathogens (1-100 CFU) can be present in large amount of food (often 1-100 g used for food analysis) while only a small volume size (10-500 µl) is used as input for the detection technique. The pathogen may also be non-uniformly distributed in the food product. Although some detection techniques have the intrinsic capacity to detect as little as one cell, the fact the detection method as such involves multiple steps starting from the food sampling to the actual test result may increase the actual detection limit significantly.

1.3.1 Low numbers

Technological interventions to preserve food in the course of its production, processing and storage together with the principles of GMP and HACCP result in the presence of low numbers of pathogens in food. The food industry uses a variety of preservation treatments, to achieve the desired product shelf-life, such as salting, drying, acidification, cold storage and freezing, heat or non thermal treatments such as high hydrostatic pressure (HHP), or intensive light pulses (ILP) or a combination of treatments etc. Some food processing employs procedures that are primary bacterio-static, other are bactericidal. Depending on the type of technological intervention and its magnitude, this intervention can be lethal or at least growth inhibitory to many of the microorganism present including the pathogens (Corbo *et al.*, 2009; Hendrickx *et al.*, 1995; Leistner and Gorris, 1995; Raso and Barbosa-Canovas, 2003).

The detection limit of the conventional isolation media and most of the alternative rapid techniques varies between 10^2 and 10^4 CFU/ml. In the case one *E. coli* O157 cell is present in 25 gram of beef, an increase from the initial contamination level should occur before a detection technique is applied to avoid false negative results.

1.3.2 Sub-lethally injured cells

However not all cells are alive or dead. Kell *et al.* (1998) described different physiological stages of microorganisms including injured cells. Food processing and preservation as described above, are also known to adversely affect microorganisms, resulting in either death, but also injury (Bautista *et al.*, 1998; Chang *et al.*, 2003; Dykes and Withers, 1999; Moorhead and Dykes, 2002; Restaino *et al.*, 2001). Despite the different technological interventions it should be noticed that the natural occurrence of injured microorganisms on contaminated foods is already long time well recognized (Busta, 1976) as most of the foods itself present a non optimal environment for microorganisms due to its intrinsic factors such as water activity, pH, etc (Wesche *et al.*, 2009).

Moreover, a sub-lethally injured cell can be defined as a cell that survives a stress, but loses some of its distinctive qualities. In general microbial sub-lethal injury is characterised by the capability of a microorganism to return to normalcy during a recovery process in which the damaged essential components are repaired. Ray and Speck (Ray and Speck, 1972) were among the first researchers to define injured cells as those capable of forming colonies on non-selective media but not on selective media. Nowadays the same definition is still handled (Besse *et al.*, 2000; Wu, 2008; Wuytack *et al.*, 2003). Additionally sub-lethally injured cells may show an extended lag phase compared with non injured cells, to repair and synthesize the proteins and the nucleic acids needed for growth (Adams, 2005; Li *et al.*, 2006; Perni *et al.*, 2007). The cell membrane appears to be the component most commonly affected (Dykes, 1999; Hurst, 1977).

The conventional selective isolation media used in food diagnostics contain a variety of selective ingredients at varying concentrations, e.g. bile salts, antibiotics such as novobiocin, acriflavin, cefsulodin, etc, that are potentially harmful to injured cells (Adams, 2005; Restaino *et al.*, 2001). Total numbers may be underestimated as sub-lethally injured cells might fail to grow, or at least show retarded growth when counted directly (Bozoglu *et al.*, 2004). Besides counting, the risk exist that sub-lethally injured cells are not able to grow in selective enrichment broth, and therefore not able to reach the detection limit of a certain detection technique, causing false negative results (Hara-Kudo *et al.*, 2000; Ngutter and Donnelly, 2003; Restaino *et al.*, 2001).

1.3.3 Matrix effect

Food components such as some kinds of proteins and fat are often described to inhibit the PCR reaction, resulting in potential false negatives results (Malorny *et al.*, 2003b; Rodriguez-Lazaro *et al.*, 2007; Rodriguez-Lazaro and Hernandez, 2006). PCR inhibitors are found in just about any food type including meat, milk, cheese, produce and spices (Wilson, 1997). For example, milk contains high levels of cations (Ca^{2+}), proteases, nucleases, fatty acids and DNA. In addition heme, bile salts, fatty acids and collagen are PCR inhibitors that are present in meat or liver (Maurer J., 2006). An appropriate sample preparation is needed to eliminate the presence of these interfering or inhibitory compounds.

Also the presence of large numbers of harmless background flora in a complex sample matrix makes it difficult to detect the presence of low numbers of food-borne pathogens. The background/accompanying flora is correlated with the food type. For example, high numbers of accompanying flora can be found in fermented foods, while less accompanying flora will be present in pasteurized foods. Even with improved methods for detecting pathogens in foods and environmental samples, microbiologists often face a “needle-in-a-haystack” challenge (Jaykus, 2003).

1.3.4 Small input volumes

The input volume used for the detection of pathogens from food based on chromogenic agars is a loop inoculation (approximately 10 µl). The input volumes for immunological based methods, such as LFD, ELISA and ELFA (VIDAS®) range from approximately 10 µl to 500 µl. Also PCR uses small input volumes in the range of 0.1 µl - 1 ml. Together with the fact that most of the time low numbers of pathogenic microorganism are present in food, an appropriate increase in the initial contamination level (growth of target microorganisms) of the food is needed for reliable detection of pathogens.

1.3.5 Conclusions

Some of the hurdles, such as sub-lethal injury and low numbers, are specifically associated with the pathogens, while others like accompanying flora, food debris and inhibitory compounds are associated with the food itself. These bottlenecks can be abated by an appropriate sample preparation. The goal of every sample preparation technique is to obtain a sample of extremely small volume in which viable target bacteria are concentrated without interfering compounds.

1.4 Sample preparation

1.4.1 Resuscitation

Sub-lethally injured cells have the capability to repair themselves and return to a normal physiological state with initiation of growth and cells division under favourable conditions. The recovery of the lost capabilities in injured cells has been termed “resuscitation”. The repair of cell ribosomes and membrane appears to be essential. Generally most cells repair within 2-4 h at a suitable incubation temperature in a nutrition rich non-selective medium. Afterwards growth can take place (Dykes, 1999; Wu, 2008).

It has been shown that enrichment broths differ significantly in the ability to resuscitate injured cells (Osborne and Bremer, 2002; Uyttendaele *et al.*, 2001a). For example the fastest recovery of acid stressed *Salmonella* occurred in universal pre-enrichment broth (Liao and Fett, 2005), freeze and heat stressed *E. coli* O157 recovered best in TSB (non-selective broth) or BCM® *E. coli* enrichment broth (selective broth) (Restaino *et al.*, 2001) while yeast extract has shown to improve the recovery of heat injured *L. monocytogenes* cells (Busch and Donnelly, 1992). A great deal of efforts has been made to develop optimal media and conditions for the recovery of bacterial pathogens from foods (Ray 1979; Mackey 2000). In general, it is recommended that samples are first incubated in pre-enrichment broth (resuscitation broth) to allow injured bacteria present to resuscitate before selective enrichment is applied (Hara-Kudo *et al.*, 2000; Rijpens and Herman, 2004).

1.4.2 Enrichment

The resuscitation step is followed by an enrichment step. Enrichment is the phase in a detection process that allows the (resuscitated) healthy target cells to multiply until detectable numbers. Therefore selective components may be added to the liquid pre-enrichment broth used for resuscitation or another more selective enrichment broth can be used. This is needed to ensure that the target cells are able to multiply under these circumstances, avoiding accompanying flora to grow. The enrichment phase is mostly an overnight or 24 h process (Gracias and McKillip, 2004).

Overgrowth of target microorganisms during enrichment, causing false negative results, is often described but mostly for taxonomic related spp. e.g. *Listeria innocua* and *L. monocytogenes* competition (Besse *et al.*, 2005; Cornu *et al.*, 2002; Flanders *et al.*, 1995; Grant, 2004). Competitor microorganisms can also interfere with and reduce the sensitivity of rapid methods. Closely related competitor organisms can cross-react with detection systems producing false positive results. Competitor cross reaction are often described with immuno-based detection methods and are the reason for the need of confirmation (to avoid false positive results).

On the one hand appropriate enrichment conditions need to favour resuscitation of sub-lethal injured cells meaning as less as possible selective agents should be present in the enrichment broth because these agents inhibit fast recovery. Especially low numbers (<100 cells) of sub-lethal injured cells, as often encountered in naturally contaminated foods, show a wide distribution of lag phase duration (Francois *et al.*, 2006; Stephens *et al.*, 1997) and may not reach the threshold level for subsequent detection with a rapid technique using a shortened enrichment procedure. On the other hand, the target microorganism may be overgrown by accompanying flora, and therefore defined amounts of selective agents are recommended (Besse *et al.*, 2005; Johansson, 1998).

1.4.3 Separation and concentration

Separation can be defined as a sample preparation that removes a select population from a complex mixture, while concentration can be defined as a sample preparation that seeks to reduce sample volume while simultaneously recovering all of the initial bacterial population of interest. Methods of bacterial separation and concentration can be categorized as chemical, physical or biological approaches, keeping in mind that many methods actually combine these general principals. In this paragraph, a summary is made of the most important techniques used in food microbiology. For detailed reading reference is made towards different review papers published earlier (Benoit and Donahue, 2003; Brehm-Stecher *et al.*, 2009; Patel, 2000; Stevens and Jaykus, 2004).

1.4.3.1 Physical methods

- *Centrifugation*

Centrifugation is a separation method that uses rotation about a fixed axis to produce the centrifugal force. This force propels particles suspended in a liquid medium to sediment, with the rate of sedimentation dependent upon a variety of physical factors. The Stokes equation can be used to describe the settling of cells in a centrifugal field such that the rate of sedimentation depends on the particle diameter, particle density, solution density, volume, angle and speed of rotation. The sedimentation rate is zero when the particle density is the same as the liquid density. The sedimentation rate decreases as the liquid viscosity increases and increases as the gravitational force increases. By manipulating variables such as solution density and particle size, alternative centrifugation methods such as density gradient and differential centrifugation (buoyant density) have been developed (Fukushima *et al.*, 2007; Lindqvist *et al.*, 1997; Lindqvist, 1997; Patel, 2000; Stevens and Jaykus, 2004; Uyttendaele *et al.*, 1999).

- *Filtration*

During filtration, a food-product or a food-product homogenate is passed through a filter and microorganisms are retained on the filter while the surrounding food sample filtrate is discarded. If desired, the organisms can theoretically be released from the filter using principles of elution. Filter type, pore shape and pore dimensions all contribute to the ability to elute microorganisms from the filter. Alternatively, filters of extremely large pore size (such as cheesecloth or filter bags) can be used to trap particulates while the bacteria flow through the filter and are recovered with the filtrate. In this case, assuring that the bacteria do not bind to the discarded particulates is critical (Stevens and Jaykus, 2004).

- *Ultrasound*

When cells in suspension are placed in an ultrasonic standing wave, the cells experience forces that move them to preferred regions within the field at half wavelength intervals. The cells also experience ultrasonically induced attractive or repulsive inter-particle forces. These forces and the subsequent movement of cells in suspension lead to the formation of high-concentration cell clusters. Relative motion between the fluid and the field or electrical modulation can be used to separate or fractionate the cell clusters from the suspension. No specific information on the ability of ultrasound to lower the bacterial detection limit of a particular detection method was reported (Benoit and Donahue, 2003).

1.4.3.2 Biological methods

- *Immunomagnetic separation and concentration*

Superparamagnetic particles (i.e. particles that are magnetized only in the presence of a magnetic field, and which demagnetize and are readily resuspended upon removal of the field) can be coated with antibodies, allowing specific capture and isolation of intact cells directly from a complex sample suspension, without the need for column immobilization or centrifugation. This method is now generically referred to as immunomagnetic separation and concentration (IMS) and nowadays widely used. Often IMS is coupled with ELISA so selective enrichment is reduced towards 24 h or coupled with PCR as IMS can aid in the elimination of inhibiting compounds (Chandler *et al.*, 2001; Chapman *et al.*, 1997). The efficiency of IMS is dependent upon the specific monoclonal antibody, the recovery procedure and sample matrix interference. Monosized superparamagnetic beads known as "Dynabeads" are available commercially from Invitrogen-Dynal (Oslo, Norway) or others such as LabM (Bury, UK). An automated system, called Pathatrix® (Matrix MicroScience Inc.) is also used nowadays. Pathatrix® is a patented re-circulating immunomagnetic separation technology. By recirculating the sample over a capture phase, comprised of immobilized antibody coated magnetic beads, the sensitivity of the capture is increased. A high volume wash enables the efficient removal of the sample matrices, non-specific microorganisms and PCR inhibitors. This technique is approved by AOAC to perform as described. IMS obtains an approximate 100-fold to 1000-fold enhancement in sensitivity over the comparative method for foods with high levels of competitive microflora (Bolton *et al.*, 1996; Call *et al.*, 2001; Chapman *et al.*, 1997; Seo *et al.*, 1998; Weagant *et al.*, 1995; Wright *et al.*, 1994).

Moreover the size and the surface of the coated particle influence the efficiency of the IMS. Therefore interest raised in the incorporation of nanoparticles (nano-biorecognition). Coating biomolecules on nanoparticles allows multiple contacts between nanomaterials and target cells, and therefore the functionalized nanomaterials display higher binding affinity than free biomolecules or biomolecules attached to superparamagnetic beads of bigger size. The incorporation of bio-functionalized nanomaterials into current pathogen detection methods has led to rapid and nearly real time pathogen detection (as short as a few minutes), improved sensitivity (single bacterial cell) and also simultaneous detection of multiple microorganisms. The development of practical biosensors using nanomaterials is promising in eliminating the need for expensive or complicated instruments and allowing the rapid detection of food-borne pathogens on a portable or hand-held device (Luo and Stutzenberger, 2008; Yang *et al.*, 2008).

- *Converted immunoassays to include bacteriophage proteins*

Bacteriophages, more in particular specific tail-associated proteins, have the potential to selectively separate and concentrate bacterial cells (Kretzer *et al.*, 2007; Loessner *et al.*, 2002). The specific tail-associated proteins can be attached to paramagnetic beads to capture bacteria in suspension. The bacteria-bead complex can be used for plating on selective chromogenic agar plates or can be used for detection with e.g. PCR (Bennett *et al.*, 1997; Favrin *et al.*, 2003; Gray and Bhunia, 2005).

Paramagnetic beads coated with phage proteins showed improved performance compared to commercially available antibody-based beads, both with respect to sensitivity and recovery (Kretzer *et al.*, 2007; Loessner *et al.*, 2002). Recently BioMérieux launched a 6-hour test, VIDAS® UP for *E. coli* O157:H7, using the recombinant bacteriophage technology. This is an automated qualitative test for the detection of *E. coli* O157 in food, feed, environmental samples and soil. The technology uses bacteriophages, special binding proteins to capture and isolate bacteria from a sample (Goodridge and Griffiths, 2002; Hagens and Loessner, 2007; Hagens and Offerhaus, 2008; Rees and Dodd, 2006; Stewart *et al.*, 1998).

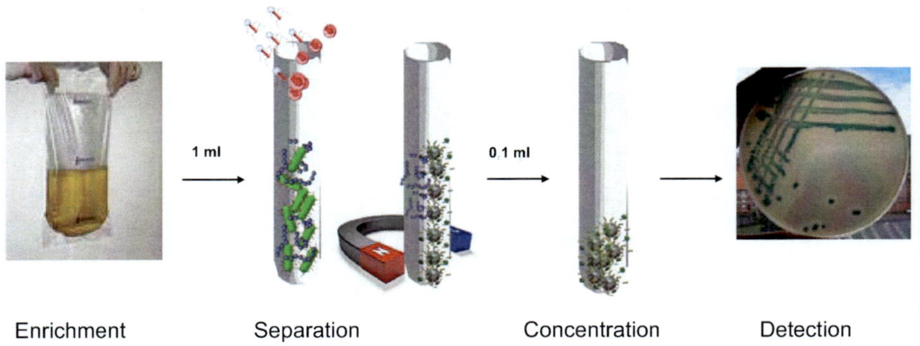

Figure 1.8. The principle of magnetic capture based on bacteriophage proteins.

1.4.3.3 Chemical methods

In the particular case of PCR based detection methods, chemical separation/concentration methods are described. These chemical separation/concentration methods are also referred to as "direct extraction" methods (Herman *et al.*, 1995; Nogva *et al.*, 2000; Ramesh *et al.*, 2002; Starbuck *et al.*, 1992). Herman *et al.* (1995) developed a direct detection method for *L. monocytogenes* in raw milk on the basis of chemical extraction of the milk components. Because of the toxicity of the reagents applied, working in an exhausting hood with protective gloves is recommended. Herman *et al.* (1995) were able to detect almost all *L. monocytogenes* cells present by this method, but these types of concentration/separation techniques are too laborious and too expensive to be applied in routine analysis (Rossmanith *et al.*, 2007).

1.4.4 Processing controls

The robustness of a method can be monitored by the use of positive and negative test controls. A processing positive control is a negative sample spiked with a sufficient amount of the target pathogen and processed through the entire protocol. A positive signal should be obtained indicating that the entire process was correctly performed. A processing negative control is a negative sample spiked with a sufficient amount of a closely related, but non-target strain, processed throughout the entire protocol. A negative signal should be obtained indicating the lack of contamination along the entire process. For this purpose, the availability reference materials and, in the particular case of PCR, of certified DNA reference materials from a reference institution could be very helpful (Malorny *et al.*, 2003a; Malorny *et al.*, 2003b; Rodriguez-Lazaro *et al.*, 2007).

Although any detection technique can produce false negative and false positive results, in the particular case of the PCR, an internal amplification control (IAC) included in every test tube, is recommended to correctly interpret the results. An internal IAC is a non-target nucleic acid sequence which is co-amplified simultaneously with the target sequence (Cone *et al.*, 1992). In a reaction with an IAC, a control signal should always be produced even when there is no target sequence present. When no control signal is observed, the reaction has failed and it indicates inhibition (Hoorfar *et al.*, 2003; Rodriguez-Lazaro *et al.*, 2007).

1.5 Conclusion of literature review

Conventional culture methods are technically simple, sensitive and capable of reliable detection of defined microorganisms. Their principal drawback is that they are labour-intensive and time-consuming. However, their ease of use, and the historical reliance of microbiologists on culture-based methods, has spawned a tradition in which they are still widely used in analytical laboratories and continue to provide a standard against which to benchmark (validate) newer techniques. Table 1.4 provides a simplified overview of some intrinsic characteristics of newer techniques used in food microbiology and their potential advantage for the user.

Molecular and immunological techniques, but also new microscopic techniques such as FISH and flowcytometry, found their way into food microbiology in the 80's and the 90's. These techniques were introduced to fasten up the detection protocol and to lower the labour-intensity of the conventional methods.

Table 1.4. Intrinsic characteristics of newer techniques as such and conventional methods.

Characteristic	Molecular techniques	Immunological techniques	Microscopic techniques	Conventional methods
Specificity based upon	DNA or RNA sequences ☺	Antibody-antigen binding, recently also phage ligands ☹/⊜	Staining using physiological or taxonomic probes ☺/⊜	Phenotypical tests (morphology, biochemical reactions, etc.) ☹/⊜
Expression of sensitivity	1 genome/PCR reaction → 10^2-10^3 CFU/ml ☺	10^4-10^5 CFU/ml ⊜	1 cell/visual field → ± 10^4 CFU/ml ⊜/☺	1 CFU/25g ☺/⊜
Time to result	1-3 h ☺	10 min - 1h ☺	5-30 min ☺	From 3-7 days ☹
Differentiation of dead/viable cells	☹/⊜	⊜	⊜/☺	☺
Interference of food sample	☹	☺	☹	☺
Simultaneous multiparameter testing	☺	⊜	⊜	☹
Reproducibility	☺	☺	☺	⊜
Potential for automation	☺	☺	⊜	☹
Labour-intensity	⊜	⊜	⊜	☹
Cost of materials	☹	⊜	☹	☺
Investment of equipment/ infrastructure	☹	⊜	☹	☺

☹: negative intrinsic characteristic/disadvantage for the user

⊜: neutral intrinsic characteristic/no explicit disadvantage or advantage for the user

☺: positive intrinsic characteristics/advantage for the user

The implementation of these techniques in food microbiology revealed some drawbacks compared to the capacity of the technique as such. The intrinsic sensitivity of molecular techniques is a positive characteristic (e.g. 1 genome/PCR reaction), but due to the needed sample preparation (e.g. PCR inhibition) and small input volume used for this technique, the sensitivity of molecular based methods in food microbiology showed to be 10^2-10^3 CFU/ml. The time to result of immunological methods is intrinsic very fast (10 min to 1 h), but due to the low intrinsic sensitivity

(10^4-10^5 CFU/ml), adequate enrichment is needed which significantly increased the time to detection of this technique in food microbiology. Therefore the judgment of a technique as such is insufficient to provide correct characteristics of the method in food microbiology. A detection method should be seen as a whole procedure including sample preparation and the detection technique as such. A reliable detection protocol does not only encompass the selection of an appropriate detection technique, but it is also largely impacted by factors such as type of food and type of microorganism. Also care should be taken about prejudices such as "results from immunological methods are less specific than results from molecular methods" as validation studies, evaluating complete detection protocols with optimized techniques (including sample preparation) show differently.

Despite these bottlenecks, it is important to realise that in addition the presence of sub-lethally injured cells can put a spoke in the wheels. Appropriate resuscitation and enrichment steps can avoid false negative results due to not reaching the needed concentration of target to have a positive result with a certain detection technique. It is therefore of the utmost importance to incorporate sub-lethally injured cells for testing performance of resuscitation and enrichment media.

Chapter 2

Establishment of procedures provoking sub-lethal injury of *Listeria monocytogenes*, *Campylobacter jejuni* and *Escherichia coli* O157 to serve method performance testing

Redrafted from

Vicky Jasson, Mieke Uyttendaele, Andreja Rajkovic and Johan Debevere

(2007)

International Journal of Food Microbiology, 118, 241-249

Chapter 2 Establishment of procedures provoking sub-lethal injury of *Listeria monocytogenes*, *Campylobacter jejuni* and *Escherichia coli* O157 to serve method performance testing

2.1 Abstract

In this chapter procedures provoking sub-lethal injury for three different pathogens are described which may be used in determination of accuracy and robustness of methods, comparison studies and or validation of rapid detection methods. Three common food-borne pathogens were used, *Listeria monocytogenes*, *Campylobacter jejuni* and *Escherichia coli* O157. The pathogens were exposed to heat stress, cold stress, freeze stress, acid stress, oxidative stress and "food" stress. Sub-lethal injury was determined by plating in parallel on selective and non-selective media. The statistical significant differences in enumeration were established. The choice of stress to create sub-lethal injury to cells depended on the fact that the procedure must be easy to handle, repeatable and relevant for stress conditions in foods, but also on the microorganism itself.

2.2 Introduction

Many types of physical or chemical treatments are used to eliminate or inhibit the growth of microorganisms in foods. Food processing often involves the use of acidulants, additives, refrigeration, heating or freezing to control microbial proliferation (Rahman, 1999). In contrast to fully viable pathogenic bacteria used for artificial contamination, bacterial cells in naturally contaminated foods are frequently impaired by sub-lethal injury as a result of having been exposed to adverse conditions of food processing. In the injured state, bacteria become sensitive to agents to which they would otherwise show resistance to (Kurbanoglu and Algur, 2006; Osmanagaoglu, 2005).

The conventional selective media used in food diagnostics contain a variety of selective ingredients at varying concentrations, e.g. bile salts, novobiocin, acriflavin, cefsulodin, etc., that are potentially harmful to injured cells (Adams, 2005; Restaino *et al.*, 2001). Total numbers are underestimated because injured cells fail to grow, or at least show retarded growth due to a longer lag phase when cultured in appropriate selective culture media. Therefore such injured cells may pose problems of detection or enumeration. It has been shown that enrichment broths differ significantly in the ability to detect injured cells (Osborne and Bremer, 2002; Restaino *et al.*, 2001; Uyttendaele *et al.*, 2001a). However, injured organisms may be capable of repair in certain food products and may possess the potential for pathogenicity, thus posing a hazard to health (Bozoglu *et al.*, 2004; Vermeiren *et al.*, 2006). Overlooking the need for resuscitation may result in a significant risk of false negative results or underestimation of the contamination level.

Whenever possible, naturally-contaminated samples should be used in any comparison or validation of microbial enumeration and detection methods (Andrews, 1996). However, for pathogenic bacteria it may not be possible to acquire a sufficient number of naturally-contaminated foods for each of the categories relevant to the validation study. Where needed, a protocol for the preparation of stressed (sub-lethally injured) organisms should be established. Reference material (RM) consisting of spray dried milk powder produced from artificially contaminated milk is available for some microorganisms. This type of RM is also included in collaborative studies in the framework of validation of test methods (In 't Veld *et al.*, 1993; In 't Veld, 1998; In 't Veld and Notermans, 1992; Janning *et al.*, 1995; Schulten *et al.*, 2000) or because of its microbiological stability serves quality control purposes. However, spray dried material only represents one type of stress which might be representative for microorganisms being present in dried foods e.g. *Salmonella* in milk powder or feed. The pathogens under consideration (*Listeria monocytogenes*, *Escherichia coli* O157 and *Campylobacter jejuni*) are however usually not associated with these types of dried foods. These three emerging pathogens have high priority in the control of the food chain, but all three are characterized with very low infectious dose (10-1000 CFU) (Jordan and Davies, 2001; Roberts *et al.*, 1996; Tilden *et al.*, 1996) and severe illness implying the importance of sensitive detection techniques (Kothary and Babu, 2001). All three pathogens have different ecological niches and growth characteristics. *C. jejuni* is a zoonotic agent, mainly associated with poultry products, and continues to be the major cause of bacterial gastrointestinal infections in humans worldwide (Murphy *et al.*, 2006). Shiga toxin producing *E. coli*, particularly strains of the O157 serogroup, zoonotic agents mainly associated with bovine products, have emerged as important food-borne pathogens (Duffy and Garvey, 2000). *L. monocytogenes*, a psychrotrophic microorganism, often present due to environmental contamination, remains a food safety issue especially in refrigerated ready-to-eat foods with extended shelf-life (Walls, 2005).

It was the objective of the present study to determine a standardized procedure of provoking injury allowing comparison of different detection and resuscitation methods, as well as, validation of advanced molecular or instrumental methods for rapid detection of food-borne pathogens. In addition it is of interest to compare the susceptibility of these three pathogens to similar type of stress conditions as encountered in food processing/food preservation.

2.3 Materials and methods

2.3.1 Bacterial strains and culture conditions

A list of the bacterial strains used in this study is shown in Table 2.1. A reference stock of each *C. jejuni* strain was kept at−70°C in horse blood (E&O Laboratories,

Bonnybrigde, England). A stock culture was kept at 4°C in Bolton broth (Oxoid, Basingstoke, England) containing the Bolton antibiotic supplement (Oxoid) in a micro-aerobic atmosphere using Campygen packs in anaerobic jars (Oxoid). The stock culture was renewed on a monthly basis. A working culture of *C. jejuni* was prepared by inoculation of 0.1 ml of the stock culture in 10 ml of fresh Bolton broth with the Bolton antibiotic supplement, and incubated for 24 h at 37°C under a micro-aerobic atmosphere.

A reference stock of each *L. monocytogenes* and *E. coli* O157 strain was kept at −70°C in Tryptone Soy Broth (TSB; Oxoid), supplemented with 0.6% yeast extract (YE; Oxoid), with 15% glycerol (Prolabo, Heverlee, Belgium). A stock culture was kept at 4°C on a Tryptone Soy Agar (TSA; Oxoid) slope supplemented with 0.6% YE. The stock culture was transferred monthly. Working cultures of *L. monocytogenes* and *E. coli* O157 were prepared by loop inoculation in 10 ml of fresh TSBYE and incubated for 24 h at 30°C or 37°C respectively.

During experimental setup, stationary phase cells were used and an initial inoculation level of approximately 10^6 bacteria per ml was obtained by appropriate ten-fold dilutions of the working culture in TSBYE for *L. monocytogenes* and *E. coli* O157 and in Bolton broth for *C. jejuni*. In order to establish the exact inoculation level, enumerations were performed by plating in parallel on the respectively selective and non-selective media of the pathogen involved. As such the performance of the selective media compared to non-selective media to support growth of healthy cells was determined.

Two strains of each pathogen were used to screen for the most appropriate condition to create sub-lethal injury (*L. monocytogenes* LFMFP 394, *L. monocytogenes* LFMFP 447, *C. jejuni* LFMFP 595, *C. jejuni* LFMFP 592, *E. coli* O157 LFMFP 326 and *E. coli* O157 LFMFP 454), additional more strains were subjected to selected stress conditions to establish strain variation (Table 2.1).

Table 2.1. Overview of the bacterial strains.

Pathogen	* LFMFP number	Origin
Listeria monocytogenes	394	Wijnendaele cheese
	447	Jalisco cheese
	034	Soft cheese
	182	Human
	235	Cooked meat
	045	Cooked pitta meat
Campylobacter jejuni	595	Bovine
	592	Poultry
	597	Water
	601	Human
	598	Water
	591	Poultry
Escherichia coli O157	454	Bovine
	326	Human
	330	Human
	328	Human
	338	Bovine
	335	Bovine
	463	**NA
	474	**NA

*** Laboratory of Food Microbiology and Food Preservation**

**** Not available**

2.3.2 Definition of sub-lethal injury

The definition of sub-lethal injury is based on an article of Besse *et al.* (2000). Sub-lethal injured bacteria become sensitive to many selective compounds due to damage in their membrane and modifications of their permeability, and they lose their ability to grow on selective media (Besse *et al.*, 2000; Besse, 2002). As such sub-lethal cells are defined as cells that are able to multiply and form colonies on non-selective media but not on selective media.

$$\% \text{ sub-lethal injury} = \frac{\text{counts on non selective media - counts on selective media}}{\text{counts on non selective media}} \times 100$$

For *L. monocytogenes*, the non-selective medium is TSA supplemented with 0.6% YE (Farber and Peterkin, 1991) and two selective media are included: Agar Listeria Ottavani & Agosti (ALOA; Biolife, Milan, Italy) and TSAYE supplemented with 40 g/l NaCl (Busch and Donnelly, 1992; Dykes and Moorhead, 2001). The non-selective medium for *C. jejuni* is Colombia agar base (Oxoid) supplemented with 5% lysed horse blood (E&O Laboratories), abbreviated as CBA. The selective medium for *C. jejuni* is *Campylobacter* Blood-Free Selective Agar (Oxoid) supplemented with the modified CCDA Selective Supplement (Oxoid), shortened as mCCDA. For *E. coli* O157 the non-selective medium is TSA supplemented with 0.6% YE and the selective medium is Sorbitol MacConkey agar (Oxoid) supplemented with cefixime tellurite selective supplement (Oxoid), abbreviated as CT-SMAC. Enumerations were preformed after 48 hours incubation for *L. monocytogenes* and *C. jejuni* at 30°C and 37°C respectively and after 24 hours incubation at 37°C for *E. coli* O157.

2.3.3 Stresses

Different treatments were selected to impose various types of stress on the selected pathogens mimicking various conditions of food processing and food preservation strategies encountered by these pathogens in the associated foods (Table 2.2).

Table 2.2. An overview of the applied stresses.

Stress	Pathogen	Treatment
Heat	*L. monocytogenes*	20 min 60°C
	C. jejuni	5 min 55°C
	E. coli O157	5 min 55°C
Cold	*L. monocytogenes*	10 days 4°C
	C. jejuni	10 days 4°C
	E. coli O157	10 days 4°C
Freeze	*L. monocytogenes*	10 days -21°C
	C. jejuni	10 days -21°C
	E. coli O157	10 days -21°C
Acid	*L. monocytogenes*	24h pH 3.8 4°C
	C. jejuni	24h pH 4.3 4°C
	E. coli O157	24h pH 3.8 4°C
Oxidation	*L. monocytogenes*	1h 1000 µM H_2O_2 37°C
	C. jejuni	1h 750 µM H_2O_2 37°C
	E. coli O157	1h 750 µM H_2O_2 37°C
"Food"	*L. monocytogenes*	10 days 6.5% NaCl pH 5 4°C 0.2% acetic acid
	C. jejuni	24h 4% NaCl 4°C
	E. coli O157	10 days 13.5% NaCl pH 4.9 4°C

Figure 2.1 provides an overview of the performed experimental setup and outlines possible effects of a given stress on the pathogen. The stress conditions applied in the current research were specifically designed to cause the formation of sub-lethally injured cells.

Figure 2.1. A summary of different kinds of cells within the population after a stress treatment.

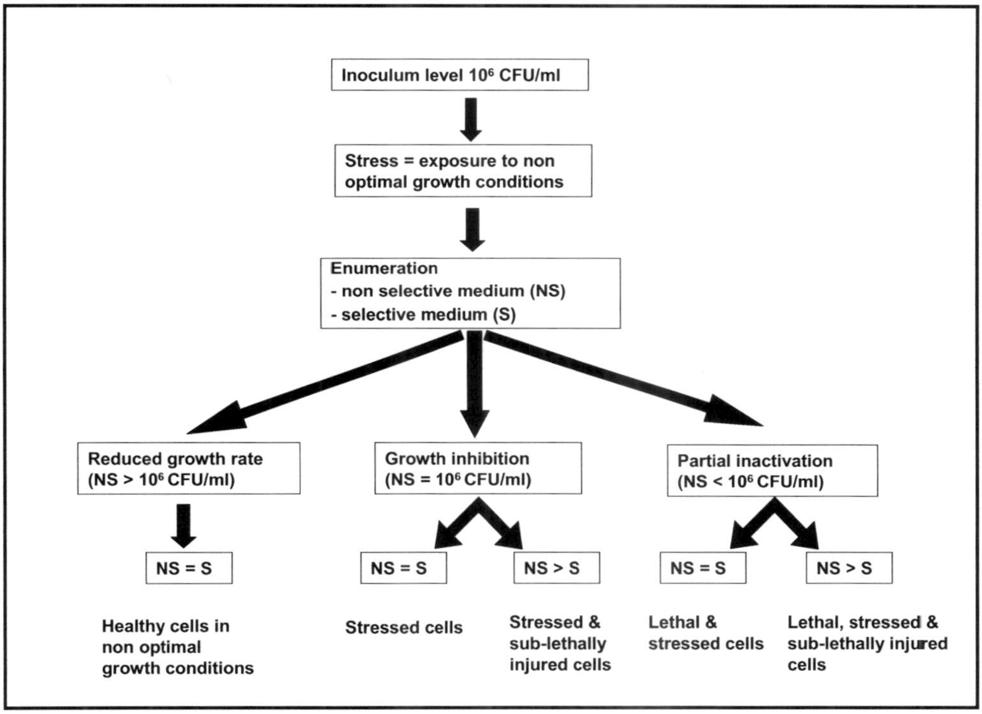

2.3.3.1 Heat stress

C. jejuni and *E. coli* O157 strains were subjected to a temperature of 55°C for 5 minutes, as suggested elsewhere (Murphy *et al.*, 2005; Uyttendaele *et al.*, 2001c). *L. monocytogenes* was subjected to a 20 minutes treatment at 60°C (Mackey *et al.*, 1994). The heating menstruum, 10 ml Bolton broth for *C. jejuni* and 10 ml TSBYE for *E. coli* O157 and *L. monocytogenes,* were preheated to test temperature prior to inoculation. After heat treatment, all the cultures were immediately put on ice and plated in parallel on selective and non-selective media.

2.3.3.2 Freeze stress

Inoculated broths were stored for 10 days in a freezer at -21°C based on Uyttendaele *et al.* (2001c). After 10 days of freeze treatment the cultures were thawed in the refrigerator at 4°C and plated in parallel on selective and non-selective media.

2.3.3.3 Cold stress

Inoculated broths were stored in a temperature controlled refrigerator at 4°C. After 10 days storage stressed cultures were plated on the respectively selective and non-selective media.

2.3.3.4 Acid stress

The acidified TSBYE for *L. monocytogenes* and *E. coli* O157 was adjusted with HCl (Merck, Darmstadt, Germany) to a pH of 3.8, filter sterilized with a 0.22 µm Cellulose Acetate Low Protein Binding Membrane (Corning Inc., New York, USA) and 10 ml distributed in sterile test tubes. Bolton broth for *C. jejuni* was adjusted to pH 4.3 with HCl based on Murphy *et al.* (2003) and then sterilized in an autoclave. The pH was checked after the sterilization. The inoculated acidified broth was kept in a refrigerator at 4°C for 24 hours upon which treated cultures were plated in parallel on the respectively selective and non-selective media.

2.3.3.5 Oxidative stress

A stock-solution of 3% hydrogen peroxide (Sigma-Aldrich, Bornem, Belgium) was made and kept in the refrigerator at 4°C for maximum 14 days. For *E. coli* O157 and *C. jejuni* 750 µM hydrogen peroxide was added in the respective cultures of each of the microorganisms and incubated for 1 hour at 37°C, based on Yamasaki *et al.* (2004). For *L. monocytogenes* 1000 µM hydrogen peroxide was added and incubated at 37°C for 1 hour (Ogawa *et al.*, 2001).

2.3.3.6 Food stress

Broths adjusted to the intrinsic factors imitating ones present in food were prepared. For *L. monocytogenes* TSBYE was supplemented with 60 g/l NaCl, adjusted with HCl to a pH of 5 and in addition 0.2 % acetic acid (Prolabo) was added to simulate the conditions of a soft cheese. The "food" broth for *E. coli* O157 was TSBYE acidified with HCl until pH 4.9 and 130 g/l NaCl was added to simulate the conditions of a fermented sausage (Stephens and Joynson, 1998). Both broths were filter sterilized. For *C. jejuni*, 35 g/l NaCl was added to Bolton Broth before autoclaving (salted minced poultry meat). The inoculated broths were kept in a temperature controlled refrigerator at 4°C for 10 days before plating on selective and non-selective media.

2.3.4 Statistical analysis

All experiments were undertaken at least in triplicate, limited by six repetitions and the results expressed are the average with the standard error of the mean shown as error bars. The significant differences between selective and non-selective media were determined with a two sided T-test using the SPSS 12.0 for Windows program (Inc., Chicago IL., USA).

2.4 Results

2.4.1 Determination of threshold values for sub-lethal injury

The three pathogens were plated on both selective and non-selective media before every treatment to verify the performance of the selective media to enumerate a non stressed culture and to determine the uncertainty of measurement that can reasonably be attributed to enumeration of microorganisms using classical plating methods (International Standards Organization, 2006a). Table 2.3 shows the p-values (T-test, α = 0.05) obtained from SPSS analysis when enumerations on selective and non-selective media before the treatment were compared (18 data sets included, i.e. 3 replicates for each of the 6 stress methods applied).

Table 2.3. The p-value (two sided T-test, $\alpha = 0.05$) of SPSS 12.0 (Inc., Chicago IL., USA) analyses selective versus non-selective media for a control culture (no treatment imposing stress).

Strain	Media	p-value	Difference
LFMFP 394	TSAYE and ALOA	0.6649	no
LFMFP 447	TSAYE and ALOA	0.8247	no
LFMFP 394	TSAYE and TSAYE-NaCl	0.1221	no
LFMFP 447	TSAYE and TSAYE-NaCl	0.1832	no
LFMFP 595	CBA and mCCDA	0.2447	no
LFMFP 592	CBA and mCCDA	0.1186	no
LFMFP 326	TSAYE and CT-SMAC	0.0069	yes
LFMFP 454	TSAYE and CT-SMAC	0.0141	yes

Table 2.3 shows that no significant difference between counts on the selective media and the non-selective medium before treatment for *L. monocytogenes* and *C. jejuni* was observed ($p > 0.05$). For *E. coli* O157 a significant difference ($p < 0.05$) in plating on CT-SMAC compared to TSAYE was observed reflecting reduced growth on the selective CT-SMAC medium. To distinguish for the relevant presence of sub-lethal cells and to compensate for inter- and intra experimental variability of measurements a threshold value was defined as percentage of sub-lethal injury from which significant level of sub-lethal injury is imposed due to treatment. The threshold values, calculated as one standard deviation above the mean of 18 data sets, were 15.3% and 29.2% for *L. monocytogenes* on ALOA and TSAYE-NaCl respectively. For *C. jejuni* and *E. coli* O157 respectively plated on mCCDA and CT-SMAC, the threshold values were 42.2% and 48.0%.

2.4.2 Determination of a procedure provoking injury

Individual cultures containing approximately 10^6 CFU/ml of each of the tested pathogens were subjected to various stresses. The effect of each stress was evaluated by post-treatment culture enumerations on selective and non-selective media. Comparison of pre- and post-treatment enumerations on non-selective media determined the extent inactivation induced by the treatment (Table 2.4).

Table 2.4. Pre- and post-stress enumerations on S and NS media of *L. monocytogenes, C. jejuni* and *E. coli* O157.

Pathogen	Type of stress	Strain	Log CFU/ml (MEAN ± stdev) Before treatment on NS medium	After treatment on NS medium	After treatment on S medium
L. monocytogenes (ALOA)	Heat	LFMFP 394	6,43 ± 0,10	3,12 ± 0,20	2,92 ± 0,12
		LFMFP 447	6,33 ± 0,09	3,10 ± 0,33	3,07 ± 0,19
	Cold	LFMFP 394	6,71 ± 0,24	8,75 ± 0,13	8,77 ± 0,13
		LFMFP 447	6,83 ± 0,35	6,11 ± 0,10	6,05 ± 0,23
	Freeze	LFMFP 394	6,18 ± 0,16	6,28 ± 0,06	6,39 ± 0,08
		LFMFP 447	6,86 ± 0,39	6,20 ± 0,14	6,10 ± 0,12
	Acid	LFMFP 394	6,64 ± 0,08	5,60 ± 0,21	4,35 ± 0,20
		LFMFP 447	6,57 ± 0,17	5,29 ± 0,16	4,85 ± 0,12
	Oxidation	LFMFP 394	6,50 ± 0,10	5,18 ± 0,13	4,96 ± 0,12
		LFMFP 447	6,37 ± 0,13	5,30 ± 0,06	5,20 ± 0,06
	"Food"	LFMFP 394	6,65 ± 0,07	6,29 ± 0,06	6,03 ± 0,13
		LFMFP 447	6,29 ± 0,10	6,29 ± 0,10	5,94 ± 0,02
L. monocytogenes (TSAYE-NaCl)	Heat	LFMFP 394	6,43 ± 0,10	3,12 ± 0,20	2,90 ± 0,14
		LFMFP 447	6,33 ± 0,09	3,10 ± 0,33	3,04 ± 0,15
	Cold	LFMFP 394	6,71 ± 0,24	8,75 ± 0,13	8,85 ± 0,07
		LFMFP 447	6,83 ± 0,35	6,11 ± 0,10	6,12 ± 0,07
	Freeze	LFMFP 394	6,18 ± 0,16	6,28 ± 0,06	6,15 ± 0,20
		LFMFP 447	6,86 ± 0,39	6,20 ± 0,14	5,97 ± 0,16
	Acid	LFMFP 394	6,64 ± 0,08	5,60 ± 0,21	3,97 ± 0,43
		LFMFP 447	6,57 ± 0,17	5,29 ± 0,16	4,72 ± 0,08
	Oxidation	LFMFP 394	6,50 ± 0,10	5,18 ± 0,13	4,42 ± 0,08
		LFMFP 447	6,37 ± 0,13	5,30 ± 0,06	4,14 ± 0,04
	"Food"	LFMFP 394	6,65 ± 0,07	6,29 ± 0,06	5,93 ± 0,07
		LFMFP 447	6,29 ± 0,10	6,29 ± 0,10	5,79 ± 0,08
C. jejuni	Heat	LFMFP 595	6,10 ± 0,13	3,41 ± 0,36	3,17 ± 0,24
		LFMFP 592	5,77 ± 0,24	2,76 ± 0,56	2,48 ± 0,31
	Cold	LFMFP 595	6,34 ± 0,10	5,60 ± 0,17	5,55 ± 0,20
		LFMFP 592	6,00 ± 0,49	5,05 ± 0,99	4,89 ± 0,85
	Freeze	LFMFP 595	5,86 ± 0,07	0,56 ± 0,31	*
		LFMFP 592	5,81 ± 0,10	1,70 ± 0,25	0,82 ± 0,34
	Acid	LFMFP 595	5,54 ± 0,11	5,02 ± 0,08	4,86 ± 0,06
		LFMFP 592	5,25 ± 0,21	5,12 ± 0,10	5,02 ± 0,08
	Oxidation	LFMFP 595	5,63 ± 0,51	3,97 ± 0,20	3,55 ± 0,31
		LFMFP 592	5,60 ± 0,34	3,65 ± 0,41	3,26 ± 0,43
	"Food"	LFMFP 595	5,50 ± 0,25	4,20 ± 0,14	3,28 ± 0,14
		LFMFP 592	5,71 ± 0,31	4,47 ± 0,15	3,55 ± 0,38
E. coli O157	Heat	LFMFP 326	6,02 ± 0,11	5,85 ± 0,06	4,96 ± 0,22
		LFMFP 454	6,17 ± 0,06	6,00 ± 0,03	5,38 ± 0,10
	Cold	LFMFP 326	6,34 ± 0,04	5,25 ± 0,11	3,98 ± 0,03
		LFMFP 454	6,20 ± 0,05	5,58 ± 0,14	4,85 ± 0,22
	Freeze	LFMFP 326	5,96 ± 0,06	5,27 ± 0,22	2,55 ± 0,34
		LFMFP 454	6,14 ± 0,01	5,39 ± 0,09	3,74 ± 0,52
	Acid	LFMFP 326	6,09 ± 0,04	6,29 ± 0,31	5,85 ± 0,07
		LFMFP 454	6,46 ± 0,18	6,12 ± 0,16	6,10 ± 0,04
	Oxidation	LFMFP 326	5,89 ± 0,07	4,25 ± 0,29	2,17 ± 0,27
		LFMFP 454	6,22 ± 0,06	4,29 ± 0,12	3,05 ± 0,24
	"Food"	LFMFP 326	5,90 ± 0,06	4,55 ± 0,36	2,24 ± 0,56
		LFMFP 454	6,10 ± 0,05	4,43 ± 0,23	2,32 ± 0,18

***: count on agar plates < 1 CFU/ml**

Almost every stress had a partial inactivation effect on the bacterial cells (Table 2.4). The only exception was found with the cold stress applied on *L. monocytogenes* LFMFP 394. The applied stress enhanced growth of the strain at 4°C providing 2 log CFU/ml increase. For *C. jejuni* the most inactivating stress was the freeze stress in contrast to *L. monocytogenes* LFMFP 447. The *E. coli* O157 strains were most susceptible to the oxidative and the "food" stress (Table 2.4).

Subsequently, the data set of results was exploited to calculate the percentage of sub-lethal injury imposed by the various stresses. Table 2.5 shows the percentage and standard deviation of the mean of sub-lethal injury for two *L. monocytogenes* strains using TSAYE and respectively ALOA and TSAYE-NaCl. Although all stresses applied (except cold stress for *L. monocytogenes* LFMFP 394) lead to inactivation for *L. monocytogenes* (Table 2.4), only the acid stress, the oxidative stress and the "food" stress induced sub-lethal injury above the threshold value.

Table 2.5 indicates strain based variation in the response of *L. monocytogenes* to the given stress. It is clear that the *L. monocytogenes* strain LFMFP 394 develops more sub-lethal cells when subjected to acid stress compared to the *L. monocytogenes* strain LFMFP 447. It can also be noted that if TSAYE-NaCl was used as the selective medium, higher percentages of sub-lethally injured cells were observed for all the stresses and both of the strains. These results indicate that the 4% NaCl supplemented TSAYE is more repressive to the target microorganism compared to the ALOA medium.

Regarding *E. coli* O157 strains all stresses tested, except the defined acid stress, were able to induce significant sub-lethal injury (Table 2.5). This is in agreement with data of (Conner and Kotrola, 1995) who reported high acid resistance of *E. coli* O157. Both *E. coli* O157 strains reacted in a similar way to the stresses applied with *E. coli* O157 LFMFP 326 being slightly more sensitive to these stress conditions than *E. coli* O157 LFMFP 454 (Table 2.4 and 2.5). Results presented in Table 2.5 show that for *C. jejuni* three stresses were able to create sub-lethal injured cells: the freezing stress, the oxidative stress and the specific "food" stress.

Table 2.5. Mean value of percentage sub-lethal injury ± standard error of the mean of *L. monocytogenes*, *E. coli* O157 and *C. jejuni* subjected to different stresses. Percentages of sub-lethal injury were subjected to One sample T-test (α = 0.05).

Stress	% sub-lethal injury							
	L. monocytogenes				*E.coli* O157		*C. jejuni*	
	ALOA [a]		TSAYE-NaCl [b]		CT-SMAC [c]		mCCDA [d]	
	LFMFP 394	LFMFP 447	LFMFP 394	LFMFP 447	LFMFP 326	LFMFP 454	LFMFP 592	LFMFP 595
Heat	31.6 ± 17.1	~ [e]	38.6 ± 6.6	2.5 ± 34.6	85.8 ± 4.8 *	75.7 ± 3.7 *	40.3 ± 21.0	53.6 ± 5.2
Cold	~ [e]	7.1 ± 20.8	~ [e]	~ [e]	94.5 ± 0.6 *	81.0 ± 2.3 *	28.1 ± 14.4	24.5 ± 27.5
Freeze	~ [e]	19.1 ± 7.3	15.6 ± 29.0	35.6 ± 19.4	99.8 ± 0.0 *	97.0 ± 1.2 *	86.4 ± 2.0 *	84.7 ± 9.7 *
Acid	92.6 ± 3.7 *	62.2 ± 1.7 *	96.1 ± 1.8 *	73.5 ± 2.2 *	30.4 ± 15.9	1.2 ± 20.3	18.9 ± 7.3	30.6 ± 4.2
Food	42.6 ± 14.8	54.5 ± 6.2 *	60.2 ± 3.7 *	67.4 ± 5.5 *	99.5 ± 0.1 *	99.2 ± 0.0 *	85.7 ± 5.3 *	87.4 ± 2.6 *
H_2O_2	38.9 ± 0.8 *	19.7 ± 2.6	82.3 ± 1.3 *	92.9 ± 0.9 *	99.1 ± 0.3 *	93.6 ± 2.1 *	59.6 ± 1.3 *	57.5 ± 12.2

a: Test value for One sample T-test = 15.3%

b: Test value for One sample T-test = 29.2%

c: Test value for One sample T-test = 48.0%

d: Test value for One sample T-test = 42.2%

e: Negative percentage of sub-lethal injury (mathematical concept)

*: Percentage sub-lethal injury is significant different (α= 0.05) from Test value

Table 2.6. Mean value of percentage sub-lethal injury ± standard error of the mean of six different *L. monocytogenes* strains subjected to acid, food and oxidative stress. Percentages of sub-lethal injury were subjected to One sample T-test ($\alpha = 0.05$).

Strain	% sub-lethal injury					
	ALOA [a]			TSAYE-NaCl [b]		
	Acid	Food	H_2O_2	Acid	Food	H_2O_2
LFMFP 394	92.6 ± 3.7 *	42.5 ± 14.8	38.9 ± 0.8 *	96.1 ± 1.8 *	60.2 ± 3.7 *	82.3 ± 1.2 *
LFMFP 447	62.2 ± 1.7 *	54.5 ± 6.2 *	19.7 ± 2.6	73.5 ± 2.2 *	67.4 ± 5.5 *	92.9 ± 0.9 *
LFMFP 182	29.2 ± 12.1	43.5 ± 7.0	68.3 ± 5.0 *	62.3 ± 8.8	40.0 ± 6.1	98.6 ± 1.1 *
LFMFP 34	13.4 ± 14.5	37.4 ± 18.0	82.0 ± 0.7 *	49.2 ± 5.8	62.8 ± 10.3	94.6 ± 4.0 *
LFMFP 45	33.5 ± 3.9 *	46.8 ± 9.3	75.3 ± 12.5 *	55.1 ± 4.1 *	56.3 ± 3.8 *	99.6 ± 0.2 *
LFMFP 235	26.6 ± 6.0	35.9 ± 5.1	82.0 ± 2.4 *	35.2 ± 2.3	38.2 ± 5.7	98.6 ± 0.2 *

a: Test value for One sample T-test = 15.3%

b: Test value for One sample T-test = 29.2%

*: Percentage sub-lethal injury is significant different ($\alpha = 0.05$) from Test value

2.4.3 Determination of strain variation

The three stresses, which created a significant percentage of sub-lethal cells, were selected to check for strain variation. The acid stress, the oxidative stress and the "food" stress were selected and applied to four additional *L. monocytogenes* strains (Table 2.6). It was noticed that in general the oxidative stress delivered the highest percentage of sub-lethal cells and that the variation in the obtained results was the smallest. The acid stress and "food" stress produced lower, but significant, percentages of sub-lethal cells and both with greater variation compared to the oxidative stress. Again TSAYE-NaCl was shown to be a more suppressing medium in comparison to ALOA enabling to establish, and more appropriate, to quantify sub-lethal injury.

Table 2.7. Mean value of percentage sub-lethal injury ± standard error of the mean of eight different *E. coli* O157 strains subjected to freeze, food and oxidative stress. Percentages of sub-lethal injury were subjected to One sample T-test (α = 0.05).

Strain	% sub-lethal injury CT-SMAC [a]		
	Freeze	Food	H_2O_2
LFMFP 454	97.0 ± 1.2 *	99.2 ± 0.0 *	93.6 ± 2.1 *
LFMFP 326	99.8 ± 0.0 *	99.5 ± 0.1 *	99.0 ± 0.3 *
LFMFP 328	98.8 ± 0.3 *	99.8 ± 0.1 *	87.0 ± 5.7 *
LFMFP 330	99.5 ± 0.5 *	99.9 ± 0.0 *	85.3 ± 3.5 *
LFMFP 335	93.6 ± 2.1 *	99.6 ± 0.1 *	77.8 ± 5.6 *
LFMFP 338	96.7 ± 0.7 *	99.4 ± 0.2 *	94.4 ± 2.3 *
LFMFP 463	99.4 ± 0.4 *	99.9 ± 0.1 *	70.3 ± 8.1
LFMFP 474	95.6 ± 1.3 *	96.0 ± 2.6 *	51.0 ± 10.6

a: Test value for One sample T-test = 48.0%

***: Percentage sub-lethal injury is significant different (α = 0.05) from Test value**

The freezing stress, the oxidative stress and the "food" stress were selected to check for strain variation with four more virulent *E. coli* O157 strains. In addition one attenuated strain of *E. coli* O157 (LFMFP 463) and one wild type *E. coli* O157 (LFMFP 474) both lacking the VT1 and VT2 genes (Vernozy-Rozand *et al.*, 2000) were also included in the experiment. Table 2.7 shows that all three stresses created consistently an amount of at least 70.3% sub-lethal injury, except for the oxidative stress in the case of the wild type *E. coli* O157 lacking the VT1 and VT2 genes (51.0%). Taking into account the threshold value, these stresses are appropriate for creating sub-lethal cells. It can be noticed that the oxidative stress shows more strain variation than the "food" stress or the freezing stress. The variation for the oxidative stress for each strain individually was also bigger compared to the "food" stress and the freezing stress.

Table 2.8 shows the results of six different *C. jejuni* strains subjected to oxidative stress and "food" stress. The results of the freezing stress are shown in Table 2.9 because of the very low numbers of survivors the calculation of the percentage sub-lethal injury was not appropriate.

Table 2.8. Mean value of percentage sub-lethal injury ± standard error of the mean of six different *C. jejuni* strains subjected to food and oxidative stress. Percentages of sub-lethal injury were subjected to One sample T-test (α = 0.05).

Strain	% sub-lethal injury mCCDA [a]	
	Food	H_2O_2
LFMFP 595	87.4 ± 2.6 *	57.5 ± 12.2
LFMFP 592	85.7 ± 5.3 *	60.0 ± 1.3 *
LFMFP 597	99.5 ± 0.1 *	97.4 ± 1.4 *
LFMFP 601	31.2 ± 8.9	63.9 ± 20.8
LFMFP 591	46.4 ± 6.8	66.6 ± 4.2 *
LFMFP 598	92.6 ± 4.3 *	63.7 ± 3.9 *

a: Test value for One sample T-test = 42.2%

***: Percentage sub-lethal injury is significant different (α = 0.05) from Test value**

Compared to *L. monocytogenes* and *E. coli* O157, *C. jejuni* showed more strain variation and more variation (also in the replicate experiments) in the percentage of sub-lethal injury imposed by the selected stresses. Although both the "food" and the oxidative stress induced sub-lethal injury, the response was not consistently significant above the threshold for all strains.

For the *C. jejuni* freeze stress the percentage of sub-lethal injury was determined after 3, 7 and 10 days, showing high susceptibility of *C. jejuni* towards freezing. After 3 days a reduction of approximately 4 log units was observed followed by a further decrease in numbers of surviving cells during the next 7 days. In addition almost all cells became sub-lethally injured (still detectable on CBA but not on mCCDA).

However, the numbers found were very low making it difficult to acquire reproducible numbers of sub-lethal injured cells.

Table 2.9. ***C. jejuni*** **strain variation tested for the freezing stress during 10 days of treatment.**

		CFU/ml					
		Day 3		Day 7		Day 10	
Strain	Repeats	CBA	mCCDA	CBA	mCCDA	CBA	mCCDA
LFMFP 591	1	190	20	20	1	1	0
	2	140	9	6	0	2	0
	3	220	11	19	0	1	0
LFMFP 592	1	370	200	120	20	95	14
	2	180	70	80	19	43	7
	3	210	60	90	13	31	3
LFMFP 595	1	200	197	87	3	8	1
	2	590	250	4	3	2	0
	3	230	140	5	4	3	1
LFMFP 597	1	210	5	47	0	5	0
	2	560	10	40	0	8	0
	3	70	2	23	3	9	0
LFMFP 598	1	70	10	0	0	5	0
	2	80	10	0	0	0	0
	3	160	10	12	0	1	0
LFMFP 601	1	140	7	0	0	0	0
	2	180	14	0	0	0	0
	3	220	20	0	0	0	0

2.5 Discussion

The purpose of the present study was to provide a comprehensive and structured examination on the effect of various food-related stress conditions on the ability to recover the food-borne pathogens, *L. monocytogenes*, *E. coli* O157 and *C. jejuni* using reference methods for enumeration including both selective and non-selective media. This was in order to set up protocols for provoking injury enabling simulations of the status of the natural flora of naturally contaminated samples. These protocols take into account the history of the cells and the degree of sub-lethal injury when assessing performance of the culturing methods.

The results showed the standard selective media for detection of *E. coli* O157 (CT-SMAC) (International Standards Organization, 2001a), to be a rather selective medium inducing already a reduction in the enumeration of cells in an optimum 24 h culture when compared to the non-selective medium. In addition upon exposure to stress conditions, except for defined acid stress, a further underestimation of the present contamination was observed which is in accordance with others (Uyttendaele *et al.*, 2001b). Also the colonies that developed on CT-SMAC were smaller compared to the TSAYE medium. These observations are in accordance with the results obtained by others (Zapparoli, 2004) who also reported that stressed cells appeared as smaller colonies.

In case of *L. monocytogenes*, ALOA seemed to be an appropriate medium for pathogen enumeration, although some underestimation of cells present was also observed with the majority of the strains if cells were exposed to some of the stress conditions (oxidative stress, "food" stress, acid stress). However, TSAYE-NaCl was shown to be more repressive for *L. monocytogenes* prohibiting the growth of injured cells both quantitatively and qualitatively to a further extent. Also here the formation of smaller colonies on TSAYE-NaCl compared to ALOA and TSAYE was noted, already for an optimum culture conditions. This is in accordance with the results of others (Jantzen *et al.*, 2006) who concluded that ALOA was a suitable medium for recover of high pressure injured sub-lethal cells. Here presented findings regarding TSAYE-NaCl remain in agreement with the various reports (Busch and Donnelly, 1992; Chawla *et al.*, 1996; Dykes and Moorhead, 2001; Flanders and Donnelly, 1994) proposing the use of TSAYE-NaCl as a "selective" (more repressive) medium to determine the fraction of sub-lethal injured *L. monocytogenes* cells upon exposure to stress.

This study showed that, although the majority of the stresses imposed created a reduction in the initial inoculum level (lethal effect) only a few stresses produced sub-lethal injured cells (as defined by a difference in recovery of cells on the selective and non-selective media). Oxidative stress induced sub-lethal injury in all three pathogens involved, although a higher (1000 µM instead of 750 µM) concentration of hydrogen peroxide was needed for *L. monocytogenes*. This can be possibly explained by the fact that Gram positive pathogens are in general more resistant to adverse conditions than for Gram negative pathogens. Oxidative stress was simple and quick (one hour) to perform making it a good candidate to be used as a standard procedure to provoke injury in a method-comparison protocol. The defined acid stress provoked sub-lethal injury for *L. monocytogenes,* but not for the generally acknowledged more acid resistant *E. coli* O157 (Law, 2000; Leyer *et al.*, 1995; Miller and Kaspar, 1994), although in both cases cells were challenged with a pH approximately 0.5 unit below their established minimum pH value for growth. On the contrary, freeze stress provoked sub-lethal injury for *E. coli* O157 and *C. jejuni,* but no significant sub-lethal injury was established for Gram positive *L. monocytogenes*. The "food" stress, a combination of addition of NaCl, reduced temperature (for *C. jejuni*) and in case of *E. coli* O157 and *L. monocytogenes* combined with reduced pH and/or acidulant was developed to simulate appropriate food conditions which each of the pathogen may encounter respectively e.g. salted minced poultry meat preparation, fermented sausage and soft cheese. The "food" stress indeed provoked a partial inactivation and created sub-lethal injured cells, a situation which is also expected to be found in naturally contaminated foods. For *C. jejuni*, it was striking that for most of the stresses involved a lethal effect was obtained, but it was not evident to obtain a protocol which provoked injury for all strains tested in this study. Strain variation and high variation in test reproducibility were noted for this pathogen confirming the capricious nature of the microorganism and illustrating more difficulties in standardization in handling of *C. jejuni* when compared to *E. coli* O157 and *L. monocytogenes*.

Stress conditions were for all three pathogens chosen to induce a subpopulation of 55 to > 90% of sub-lethal injured cells in the treated culture. This implies a possible false negative detection on selective media when low levels of pathogens are present. However, with high counts of pathogens present, even 90% of sub-lethal cells (in absolute numbers) will leave high proportion of non sub-lethally injured cells allowing detection of pathogen on the selective media. It can be noted already from the threshold value that the uncertainty of measurement of a microbiological determination is rather high. In interpretation of microbiological results of food analysis often high 95% confidence interval are to be taken into account (reproducibility limit mentioned by (International Standards Organization, 2003a) is ± 0.45 log unit). As sub-lethal injury was only taken as significant above the threshold value taking into account the intrinsic dispersion of values due to the procedure of the plating method, it is clear that the significant sub-lethal injury reported in this study will indeed lead to a significant underestimation of the contamination levels if plating is performed on selective media. Other studies (Besse *et al.*, 2000; Dykes and Withers, 1999; Hara-Kudo *et al.*, 2000; Raccach *et al.*, 2002) often relate percentage of sub-lethal injury as such without a threshold value making appropriate results interpretation difficult. With regard to lack of detection of the pathogens, and thus reporting of false negative results because of sub-lethal injury, the proportion of sub-lethal injured cells will probably not influence the presence/absence testing, if high numbers of the pathogens are present. In detection protocols using an enrichment protocol the small fraction of non-injured cells will multiply uninhibited and quickly dominate the culture. However, if presence/absence testing is performed for detection of very low numbers being present, the sub-lethal injured fraction may become very important, e.g. as noted for the freeze stress of *C. jejuni* resulting in low numbers of survivors, but practically all sub-lethal injured. Also if low numbers are present (< 100 CFU/ml) and high percentages of sub-lethal injury are noted (> 90% as noted for *E. coli* O157) the number of uninjured cells is reduced to the single cell levels and variation in lag-phase distribution becomes important as well (Francois *et al.*, 2006; Stephens *et al.*, 1997). In these cases, resuscitation of the sub-lethal injured cells will need a prolonged incubation time or a prior incubation in a non-selective medium to induce repair. This phenomenon can partly explain the failure of some (advanced) detection methods to detect low numbers of pathogens in processed foods if the enrichment protocol is too short or does not allow appropriate resuscitation of injured cells. This points out the need for including sub-lethal injured cells into the validation studies in order to asses the potential of conventional cultural methods, as well as, advanced detection methods to reliable detect and/or enumerate food-borne pathogens in foods. As naturally contaminated foods are not always available, the procedures for provoking injury established in this study for each of the three pathogens involved may aid in establishing the accuracy, the detection limit and robustness of the screening methods for pathogens in food diagnostics. Further studies are needed to investigate resuscitation mechanisms of cells injured by different sub-lethal stresses in order to have a clear overview of the capacity of a certain resuscitation broth to recover injured cells, specific for each stress.

2.6 Conclusion

This study showed that, although the majority of the stresses imposed created a reduction in the initial inoculum level only a few stresses produced sub-lethal injured cells. Oxidative stress induced sub-lethal injury in all three pathogens involved, although a higher (1000 µM instead of 750 µM) concentration of hydrogen peroxide was needed for *L. monocytogenes*. The defined acid stress provoked sub-lethal injury for *L. monocytogenes,* but not for the generally acknowledged more acid resistant *E. coli* O157. On the contrary, freeze stress provoked sub-lethal injury for *E. coli* O157 and *C. jejuni,* but no significant sub-lethal injury was established for *L. monocytogenes*. The "food" stress, a combination of addition of NaCl, reduced temperature (for *C. jejuni*) and in case of *E. coli* O157 and *L. monocytogenes* combined with reduced pH and/or acidulant was developed to simulate appropriate food conditions which each of the pathogen may encounter respectively e.g. salted minced poultry meat preparation, fermented sausage and soft cheese. The "food" stress indeed provoked a partial inactivation and created sub-lethal injured cells, a situation which is also expected to be found in naturally contaminated foods.

This study points out the need for including sub-lethal injured cells into the validation studies in order to asses the potential of conventional cultural methods, as well as, advanced detection methods to reliable detect and/or enumerate food-borne pathogens in foods. As naturally contaminated foods are not always available, the procedures for provoking injury established in this study for each of the three pathogens involved may aid in establishing the accuracy, the detection limit and robustness of the screening methods for pathogens in food diagnostics.

The procedures provoking sub-lethal injury determined in Chapter 2 are further used in Chapters 3, 4 and 5 of this PhD dissertation, to test performance of different resuscitation/enrichment broths for the involved pathogens. In Chapter 3 focus is set on the recovery of sub-lethally injured *L. monocytogenes* cells and these cells are used to determine the best enrichment broth for reliable and fast detection of low numbers of *L. monocytogenes* from foods. The alternative rapid detection and separation/concentration methods FISH and bacteriophage based capture as described in the respectively section 1.2.3.5 and 1.4.3.2 of Chapter 1 were used as rapid detection methods in Chapter 3.

Chapter 3

Kinetics of resuscitation and growth of *Listeria monocytogenes* as a tool to select appropriate enrichment conditions as a prior step to rapid detection methods

Redrafted from:

Vicky Jasson, Andreja Rajkovic, Johan Debevere and Mieke Uyttendaele

(2009)

Food Microbiology, 26, 88-93

Chapter 3 Kinetics of resuscitation and growth of *Listeria monocytogenes* as a tool to select appropriate enrichment conditions as a prior step to rapid detection methods

3.1 Abstract

Rapid methods still rely on a prior (shortened) enrichment step before application. Quantitative information is a prerequisite for understanding the resuscitation kinetics of the growth during the enrichment step. In this study various basal (non-selective) and newly introduced selective enrichment broths were evaluated. First, growth parameters (λ, μ_{max}) of both healthy and sub-lethally injured cells were determined. Next, a selection of enrichment broths was compared for their capacity to support detection within 24 h of low numbers of *Listeria monocytogenes* in artificially and naturally contaminated food samples. Detection was performed either by phage protein based capture (*Listeria* Capture kit, Profos, Regensburg, Germany) combined with plating on chromogenic medium or by fluorescence *in situ* hybridization (FISH) using the VIT-*Listeria* kit (Vermicon, Munich, Germany). Kinetics of resuscitation and growth of *L. monocytogenes* in various enrichment broths showed that for detection of low numbers of sub-lethally injured *L. monocytogenes* cells at least an overnight enrichment was needed. A selective enrichment broth was needed to enable proliferation of *L. monocytogenes* within the indigenous bacterial flora present in foods. However, combination of an appropriate enrichment condition with advanced detection techniques may enable a 24 h detection of *L. monocytogenes*.

3.2 Introduction

The widespread prevalence of *Listeria monocytogenes* (Lianou and Sofos, 2007), the severity of illness (McLauchlin *et al.*, 2004) and the number of reported outbreaks (Rocourt *et al.*, 2000) imply the importance of sensitive and rapid detection methods of this pathogen in foods. Because food, due to processing and preservation techniques (salting, acidification, freezing, packaging, etc.) presents non-optimal conditions for microorganisms, the bacterial cells are stressed and/or sub-lethally injured (Chapter 2).

Non optimal conditions influence the growth of microorganisms, either no growth or a retarded growth occurs due to the non-optimal function of the cell (Ribeiro *et al.*, 2006). When a microorganism becomes sub-lethally injured due to stress it shows a longer lag phase compared to a healthy cell (Van Houteghem *et al.*, 2008). The sub-lethal physiological state of the bacterium causes two important phenomena in food diagnostics. First a sub-lethal cell becomes sensitive to selective components in media where to which it normally shows resistance, resulting in an underestimation of the true numbers present when plating on a selective agar (Chapter 2). Secondly a sub-lethal cell shows an increased lag phase due to time needed to repair, resulting in a true risk of not reaching the detection limit of a certain detection method in the timeframe used (Deisingh and Thompson, 2004).

The detection limit of many advanced detection methods such as PCR, flow cytometry or biosensors lies in the range of 10^2-10^4 CFU/ml (Chitarra and van den Bulk, 2003; Gunasekera *et al.*, 2003; Rodriguez-Lazaro *et al.*, 2007; Tokarskyy and Marshall, 2008). However, often legislation demands presence/absence testing per 25 g or per gram (EU, 2005). Subsequently these rapid methods still rely on a prior (shortened) enrichment step before application of the immunological, molecular or microscopic advanced detection technique. Overlooking the need for optimization of the enrichment step prior to using a rapid detection method can lead to false negative results. Especially low numbers (< 100 cells) of sub-lethal injured cells, as often encountered in naturally contaminated foods, show a wide distribution of lag-phase duration (Francois *et al.*, 2006; Stephens *et al.*, 1997) and may not reach the threshold level for detection within a shortened enrichment procedure. Appropriate enrichment conditions need to favour resuscitation of sub-lethal injured cells meaning as less as possible selective agents should be present in the enrichment broth because these agents inhibit fast recovery. On the other hand, the target microorganism may be overgrown by accompanying flora, thus certain amounts of selective agents are recommended (Besse *et al.*, 2005; Johansson, 1998).

Quantitative information is a prerequisite for understanding the kinetics of resuscitation and growth during the enrichment step. It was the objective of the present study to determine, based on lag phase (λ) and maximum specific growth rate (μ_{max}) the best enrichment broth prior to rapid and reliable detection of low numbers of *L. monocytogenes* in food samples. Various basal and recently developed selective enrichment broths were evaluated. First by determination of growth parameters (λ, μ_{max}) of both healthy and sub-lethally injured cells and next by comparison of a selection of enrichment broths for their capacity to support 24 h detection of low numbers of *L. monocytogenes* in artificially and naturally contaminated food samples using either phage protein based *Listeria* Capture kit (Profos, Regensburg, Germany) combined with plating on chromogenic medium or fluorescence *in situ* hybridization (FISH) using the VIT-*Listeria* kit (Vermicon, Munich, Germany).

3.3 Materials and methods

3.3.1 Bacterial strains and culture conditions

Two *L. monocytogenes* strains (LFMFP 394 and LFMFP 447) were used and strains were cultured as described in Chapter 2 section 2.3.1.

3.3.2 Enrichment broths involved

3.3.2.1 Basal (non-selective) broths

Eight basal broths were selected: Tryptone Soy Broth (**TSB**; brand 1, brand 2 and brand 3), TSB supplemented with yeast extract (**TSBYE**; TSB (Oxoid) + 0.6% YE (Oxoid)), buffered TSB (**bTSBYE**; TSB (Oxoid) + 0.6% YE (Oxoid) + 1.35 g/l KH_2PO_4

(Fluka, Bornem, Belgium) + 9.6 g/l Na_2HPO_4 (Sigma-Aldrich, Bornem, Belgium)), Fraser base (Oxoid), Nutrient broth (**NB**; Oxoid) and *Listeria* Repair Broth abbreviated as **LRB** (Busch and Donnelly, 1992).

3.3.2.2 Selective enrichment broths

Three selective enrichment broths were selected for *L. monocytogenes*. These broths are especially designed to selectively enrich food samples with *Listeria* spp. while shortening enrichment procedure (one-step enrichment instead of two-step enrichment protocol as prescribed by ISO 11290:1996 (International Standards Organization, 1996). One Enrichment Broth (**OEB**) was obtained from Oxoid, *Listeria* Xpress Broth (**LX**) from BioMérieux (Marcy l'Etoile, France) and *Listeria* Special Broth (**LSB**) from Bio-Rad (Marnes-la-Coquette, France). Demi-Fraser broth (BioMérieux) was taken as the reference for selective pre-enrichment (International Standards Organization, 1996).

3.3.3 Determination of μ_{max} and $\lambda_{control}$ in various broths

Fresh cultures (24 h) were tenfold diluted and approximately 3.5 log CFU were inoculated (300 µl total volume) in the first well of 96 well microtiter plate (Greiner bio-one, Wemmel, Belgium). Twofold dilutions were made along the rows of the microtiter plate and triplicate experiments were performed. The initial inoculum level was confirmed for each of the first wells by plating on TSAYE and incubation for 24 h at 30°C, providing log $CFU/ml_{initial}$. Growth at 30°C was measured by optical density (OD) measurements in 10 minutes intervals using a VERSAmax™ microplate reader (Molecular Devices, Sunnyvale, CA, USA) at wavelength 380 nm except Demi-Fraser, LX, LSB and OEB, which were measured at 600 nm. For each OD reading the OD_{blank} (non-inoculated broth) was subtracted from the OD reading (resulting in value) and the average of three replicates was calculated to obtain ΔOD growth curves. For each broth, in a prior experiment (OD measurement of a twofold serial dilution of a working culture with a defined CFU/ml), the minimum number of cells needed to record a change in ΔOD reading ($\Delta OD>0$) was established resulting in log $CFU/ml_{detection}$.

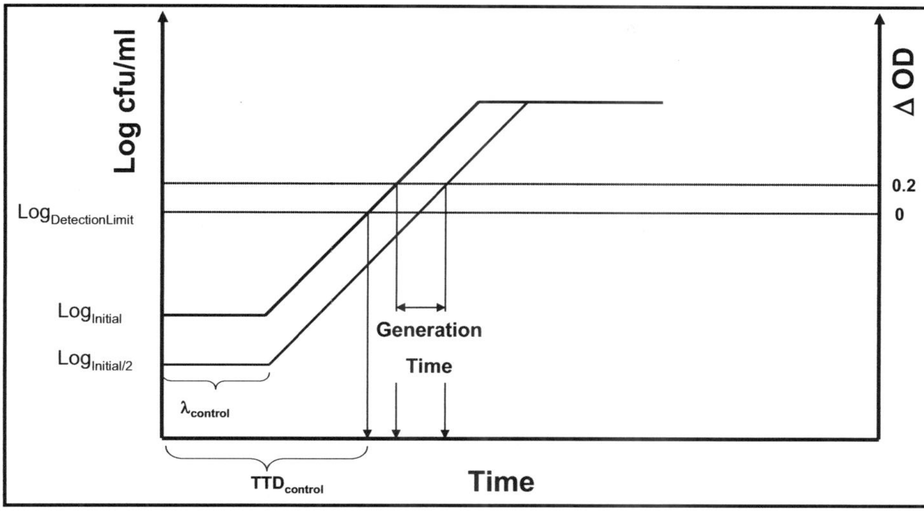

Figure 3.1. A schematic representation of the experimental setup for determination of μ_{max} of *L. monocytogenes* cells in broth ($\Delta OD = OD_{inoculated} - OD_{blank}$).

A theoretical approach on how the generation time (GT) was obtained is shown in Figure 3.1. The fixed ΔOD value was set on 0.2, except for NB and LSB (fixed ΔOD value was 0.1). The difference in time needed for two successive (twofold) dilutions to reach a fixed ΔOD was used as a measure of the GT (Figure 3.1). Data of wells inoculated with approximately 3000 and 1500 healthy cells were used to determine the GT. Low inoculum levels (< 1000 cells) were not included to eliminate the risk of encountering the variability in the length of the lag phase of individual cells (Francois *et al.*, 2006). The maximum specific growth rate (μ_{max}) was calculated as shown in Equation 3.1. Triplicate experiments were performed.

Equation 3.1

$$\mu_{max} = \frac{\log_{10}(2)}{\text{Generation Time}}$$

$\lambda_{control}$ was determined as shown in Equation 3.2. Time to detection ($TTD_{control}$) of the cells was obtained using the linear regression ($R^2 > 0.98$) and the point of ΔOD=0 (Figure 3.1 and 3.2).

Equation 3.2

$$\lambda_{control} = \frac{(\log_{10} CFU/ml_{initial} - \log_{10} CFU/ml_{detection} + \mu_{max} \times TTD_{control})}{\mu_{max}}$$

3.3.4 Determination of λ_{stress} in various broths

Three stresses capable of creating a significant percentage of sub-lethally injured *L. monocytogenes* cells were determined in Chapter 2. An acid stress (pH 3.8, adjusted with HCl, 24 h at 4°C), the "food" stress (pH 5, 0.2% acetic acid, 6.5% NaCl, 10 days at 4°C) and the oxidative stress (1000 µM H_2O_2, 1 h at 37°C) were chosen to create injured cells. After treatment approximately 3.5 log was inoculated in the wells of the first column of the microtiter plate as described above. The same suspensions of stressed cells were used for the different enrichment broths to eliminate minor variation between stressed cultures.

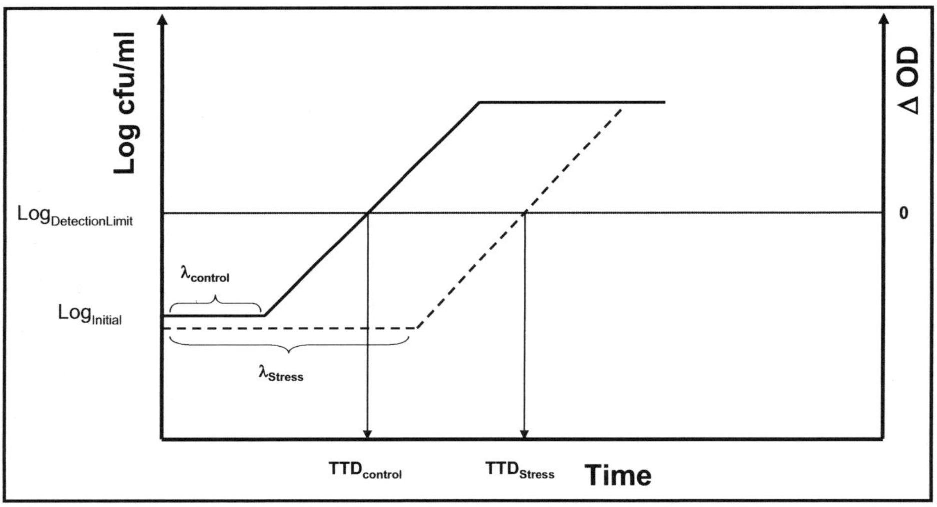

Figure 3.2. Schematic representation of the determination of λcontrol, λstress ($TTD_{control}$ = Time to detection of non stressed cells; TTD_{stress} = Time to detection of stressed cells).

Figure 3.2 explains the theoretical hypothesis of the extra time needed for sub-lethally stressed cells to reach the detection limit. The increased λ symbolized by λ_{stress} is calculated as shown in Equation 3.3. The growth rate (μ_{max}) on the other hand is not affected by the sub-lethal injury.

Equation 3.3

$$\lambda_{stress} = \frac{(\log_{10} CFU/ml_{initial} - \log_{10} CFU/ml_{detection} + \mu_{max} \times TTD_{stress})}{\mu_{max}}$$

3.3.5 Evaluation of enrichment broths using artificially contaminated samples

Food samples (10 g of endive, smoked salmon, deli salad, cheese and pâté) were artificially contaminated with 5-12 *L. monocytogenes* LFMFP 394 cells. Inoculated samples were stabilized during 17 h at 2°C followed by resuspension in 90 ml enrichment broth. TSB (brand 3), LX, LSB and Demi-Fraser were compared for their capacity to recover cells in combination with the *Listeria* Capture kit (Profos) and with VIT-*Listeria* Kit (Vermicon). Detection was achieved from 1 ml suspensions after 6 and 24 hours incubation at 30°C for Demi-Fraser and 37°C for LX, LSB and TSB.

3.3.6 Evaluation of enrichment broths using naturally contaminated samples

A variety of food samples were first screened with a routine AFNOR validated rapid (60 h) ELISA method (VIDAS® LMO II, BioMérieux) as described in section 1.2.3.2 of Chapter 1 and *L. monocytogenes* positive samples were collected and kept at -20°C (1-4 weeks) until analysis (defrosted at 2°C). 10 g portions were suspended in respectively 90 ml Demi-Fraser, LX and LSB. Samples were incubated for 17 h at 30°C (Demi-Fraser) and at 37°C (LX and LSB) before aliquots of 1 ml were taken for analyses with *Listeria* Capture Kit and VIT-*Listeria* Kit. An additional 1 ml of the sample enriched in Demi-Fraser broth was transferred to 10 ml of Fraser broth (BioMérieux), incubated for 24 h at 37°C and analysed with VIDAS® LMO II in order to confirm prior positive result obtained during screening.

3.3.7 The detection methods

3.3.7.1 *Listeria* Capture Kit

Protocol of *Listeria* Capture kit was performed as described by the manufacturer (Profos). Briefly 1 ml of the enriched sample and 20 µl of phage protein coated magnetic beads were incubated at room temperature in an overhead rotor for 20 minutes. Magnetic field was applied for 5 minutes and supernatant was carefully aspirated and discarded. Samples were washed 3 times with cell wash buffer, provided by the kit. Supernatant was discarded and the particle-bacteria complex was resuspended in 100 µl of cell wash buffer before plating on two Agar *Listeria* Ottaviani & Agosti plates (ALOA; Biolife, Milan, Italy). Counting of colony forming units was performed after 24 hours incubation at 37°C.

3.3.7.2 *Listeria* Vermicon Identification Technology kit

Protocol of VIT-*Listeria* Kit was performed as described by the manufacturer (Vermicon). Briefly 1 ml of the enriched samples was centrifuged (Centrifuge 5415C, Eppendorf, Hamburg, Germany) (room temperature, 10 minutes, 1000g) and supernatant was discarded. Three drops of each sample were added to a microscopic slide and fixation was performed as described in the protocol. Negative and positive control probes were included. Washing steps were performed and samples were evaluated microscopically under the epifluorescent microscope (Axio Imager A1, Carl Zeiss MicroImaging Inc., Thornwood, NY, USA) (100x immersion oil objective).

3.3.8 Statistical analysis

For each individual strain the obtained μ_{max} were analyzed by one-way ANOVA (SPSS, Inc., Chicago IL., USA). Post hoc test were performed using Tukey test (α=0.05). The results of the ANOVA are indicated with Tukey based homogeneous subsets (group A, B, C, D and E significant difference α=0.05).

3.4 Results

3.4.1 Determination of μ_{max}

Table 3.1 shows the μ_{max} of *L. monocytogenes* cells grown at 30°C in non selective and selective broths. Post hoc test results are indicated in Table 3.1 with different homogeneous subsets. Subset A indicates the greatest μ_{max} meaning the fastest growth of *L. monocytogenes* cells. *L. monocytogenes* LFMFP 394 inoculated in the non-selective basal broths TSB (3), TSBYE and NB resulted in fastest μ_{max}.

Table 3.1. Average μ_{max} and its standard deviation at 30°C of two different *L. monocytogenes* strains in 12 different broths.

Broth	*L. monocytogenes* LFMFP 394 μ_{max} (h^{-1})	*L. monocytogenes* LFMFP 394 Stdev	*L. monocytogenes* LFMFP 447 μ_{max} (h^{-1})	*L. monocytogenes* LFMFP 447 Stdev
TSB (1)	0.39 $^{(D,C,B)}$	0,05	0.41 $^{(B,A)}$	0,04
TSB (2)	0.36 $^{(D,C,B)}$	0,05	0.36 $^{(C,B)}$	0,01
TSB (3)	0.51 $^{(A)}$	0,06	0.42 $^{(B,A)}$	0,05
TSBYE	0.40 $^{(C,B,A)}$	0,02	0.40 $^{(B,A)}$	0,03
bTSBYE	0.32 $^{(D,C,B)}$	0,04	0.38 $^{(C,B)}$	0,04
Fraser-base	0.31 $^{(D,C,B)}$	0,03	0.32 $^{(C,B)}$	0,05
LRB	0.38 $^{(D,C,B)}$	0,06	0.35 $^{(C,B)}$	0,00
NB	0.42 $^{(B,A)}$	0,05	0.50 $^{(A)}$	0,09
Demi Fraser	0.19 $^{(E)}$	0,02	0.19 $^{(D)}$	0,03
LX	0.29 $^{(E,D)}$	0,01	0.27 $^{(D,C)}$	0,01
LSB	0.36 $^{(D,C,B)}$	0,03	0.33 $^{(C,B)}$	0,05
OEB	0.31 $^{(D,C)}$	0,01	0.31 $^{(C,B)}$	0.01

Tukey based homogeneous subsets (group A, B, C, D and E significant difference α=0.05).

For *L. monocytogenes* LFMFP 447 TSB (1), TSB (3), TSBYE and NB showed a μ_{max} which categorizes in homogeneous subset A. Significant differences can be noticed between TSBs obtained from different producers. The results clearly showed a significantly lower μ_{max} in Demi-Fraser compared to all other media tested, meaning that this (selective) broth supported a significantly slower growth of *L. monocytogenes*. With regard to the other selective enrichment broths, LSB showed overall a higher μ_{max} than Demi-Fraser, LX or OEB and was comparable to many of the non selective broths such as TSB (2), bTSBYE, LRB or Fraser base.

3.4.2 Determination of $\lambda_{control}$ and λ_{stress}

Figure 3.3 shows the results for the length of the lag phase for *L. monocytogenes* LFMFP 394 subjected to no stress (control) and after exposure to "food", acid and oxidative stress. Results for *L. monocytogenes* LFMFP 447 were similar (Figure 3.4).

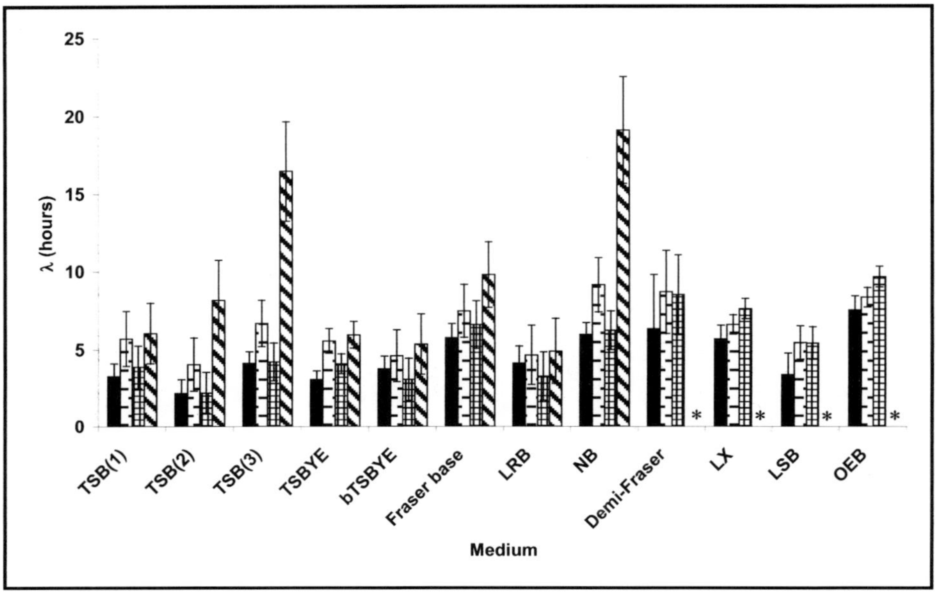

Figure 3.3. **λ of *L. monocytogenes* LFMFP 394 at 30°C. ■ represents $\lambda_{control}$. ▭ represents λ of acid stressed cells, ▦ food stressed and ▧ oxidative stressed *L. monocytogenes* LFMFP 394 cells. * no growth was detected. Error bars represent cumulative standard deviation.**

Figures 3.3 and 3.4 show a trend with an increase in lag phase when cells were subjected to prior stress conditions. It can be noticed that the LRB a broth especially designed to recover injured cells (Busch and Donnelly, 1992) did not support a significantly shorter lag-phase compared to e.g. TSBYE, except for the food and the oxidative stress applied to *L. monocytogenes* LFMFP 447. It can be concluded from Figures 3.3 and 3.4 that the oxidative stress had the most severe influence on the length of the lag phase, especially for TSB (3) and NB for which the lag-phase was prolonged from approximately 5 h to >15 h. These results indicate that it can not be taken for granted that a broth supporting a significantly faster μ_{max} for *L. monocytogenes* cells such as TSB (3) and NB is also the best to support resuscitation. In the four selective broths no recovery from oxidative stress was noticed illustrating the adverse character of selective agents to resuscitation.

Based on quantitative data from lag phase (λ) (Figure 3.3) and maximum specific growth rate (μ_{max}) (Table 3.1), TSB (1) as a non-selective broth, LX and LSB as selective broths and Demi-Fraser because it is recommended in ISO 11290:1996 (International Standards Organization, 1996) were further evaluated in a (shortened) enrichment procedure for rapid detection of low numbers of *L. monocytogenes* in foods.

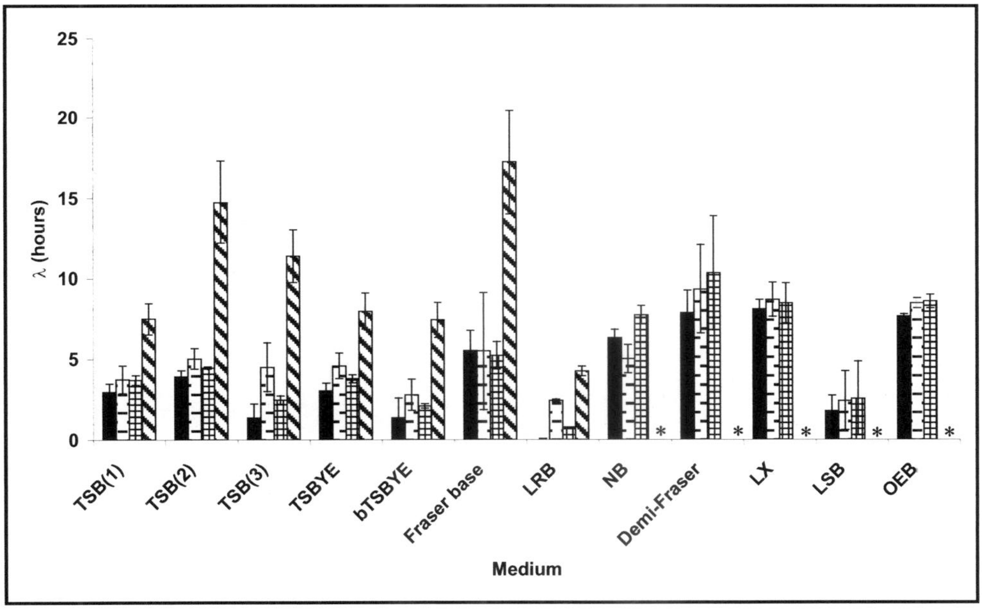

Figure 3.4. λ of *L. monocytogenes* LFMFP 447 at 30°C. ■ represents $\lambda_{control}$. ▭ represents λ of acid stressed cells, ▦ food stressed and ▧ oxidative stressed *L. monocytogenes* LFMFP 394 cells. * no growth was detected. Error bars represent cumulative standard deviation.

3.4.3 Detection of *L. monocytogenes* in artificially contaminated samples

Results are shown in Table 3.2. Deli salad was shown to be a difficult matrix for rapid detection of *L. monocytogenes*. Probably the acidity of the deli-salad (approximately pH. 4.5) and the presence of preservatives (approximately 0.7% acetic acid) seriously inhibited the growth of *L. monocytogenes* prohibiting detection after 6 hours enrichment. The selective enrichment broths LX and LSB enabled detection of low numbers of *L. monocytogenes* in the remaining 4 food matrices after 6 hours enrichment time if combined with phage protein based capture and plating.

Table 3.2. Detection of *L. monocytogenes* LFMFP 394 cells with *Listeria* Capture kit and VIT-*Listeria* Kit (VIT) in artificially contaminated food samples (5-12 CFU/g) after 6 hours and 24 hours enrichment in 4 different enrichment broths. Demi-Fraser was incubated at 30°C and the other media at 37°C.

Food sample	Incubation time (h)	Demi-Fraser Capture*	Demi-Fraser VIT**	LX Capture*	LX VIT**	LSB Capture*	LSB VIT**	TSB (1) Capture*	TSB (1) VIT**
Endive	6	+	-	+	-	+	+	-	-
	24	++++	++	++++	++++	++++	++++	++	++
Deli salad	6	-	-	-	-	-	-	-	-
	24	++	-	++	++	-	-	-	-
Cheese	6	+	++	+	++	+	+++	-	++
	24	++++	++	++++	+++	++++	++++	+	++
Smoked salmon	6	-	-	+	-	+	-	-	-
	24	+++	++	++++	++++	++++	++++	+++	-
Pâté	6	+	-	++	++	++	-	+	-
	24	++++	++	++++	+++	++++	++++	+++	++

Capture*: *Listeria* Capture Kit

VIT: VIT-*Listeria* Kit**

<u>For capture:</u>

-: absence; +:1 ≤ CFU/plate ≤ 14; ++:15 ≤ CFU/plate ≤ 150; +++: 151 ≤ CFU/plate; ++++ more than countable

<u>For VIT-*Listeria* Kit:</u>

-: absence; +:1-20 cells/10 visual fields; ++:1-20 cells/visual fields; +++: 21-100 cells/visual field

++++ more then 100 cells/visual fields

Although TSB (1) showed a significantly higher μ_{max}, detection of *L. monocytogenes* only occurred in 2/5 food matrices after 6 hours incubation, one with capture and one with VIT-*Listeria* Kit. Demi-Fraser, although having the lowest μ_{max} was capable of supporting growth enabling detection of the low numbers of pathogen in 3/5 inoculated food matrices after 6 hours. The non-selective broth failed to support growth probably due to suppression of the pathogen by competitive growth of indigenous bacteria in the food. For detection with VIT-*Listeria* Kit it can be noticed that enrichment in LX and LSB also resulted in more positive detections after 6 hours compared to Demi-Fraser and TSB. However, overall results in Table 3.2 indicate that an incubation time of 6 hours is insufficient for any of the enrichment broths to obtain detection in all 5 implied food matrices whereas the 24 hours enrichment in LX broth succeeded in detection of *L. monocytogenes* in all food samples whether using phage protein based capture with plating or VIT-*Listeria* Kit. These results are in correlation with the data presented in Figures 3.3 and 3.4 which depicted already a lag phase of approximately 3 h for healthy *L. monocytogenes* cells.

3.4.4 Detection of *L. monocytogenes* in naturally contaminated samples

Table 3.3 shows the results of *L. monocytogenes* detection with phage protein based capture combined with plating, VIT-*Listeria* Kit and VIDAS® LMO II in naturally contaminated samples after 17 hours of enrichment in 3 different selective enrichment broths.

Table 3.3. Detection of *L. monocytogenes* with *Listeria* Capture kit, VIT-*Listeria* Kit (VIT) and VIDAS® LMO II in naturally contaminated food samples after 17 hours enrichment in 3 different enrichment broths. Demi-Fraser was incubated at 30°C and the other media at 37°C.

Food sample	LX		LSB		Demi-Fraser		
	Capture*	VIT**	Capture*	VIT**	Capture*	VIT**	VIDAS®
Shrimps	+	+	-	+	+	-	2.49
Surimi	++++	++	+++	++	+++	-	2.49
Hamburger 1	++++	+	+	+	+++	-	0.81
Hamburger 2	+	+	+	-	++	-	0.78
Deli salad	++++	+++	+	+	+++	-	2.80
Endive	++++	+++	+	++++	+++	++	2.48
Cheese	+	-	+	-	++	-	2.90
Vegetables	++	++	+	+	+++	++	2.73
Cheese	++++	++	++++	++	++	-	2.88

Capture*: *Listeria* Capture Kit

VIT: VIT-*Listeria* Kit**

For capture:

-: absence; +:1≤CFU/plate≤14; ++:15≤CFU/plate≤150; +++: 151≤CFU/plate; +++* more than countable

For VIT-*Listeria* Kit:

-: absence; +:1-20 cells/10 visual fields; ++:1-20 cells/visual fields; +++: 21-100 cells/visual field

++++ more then 100 cells/visual fields

For VIDAS® LMO II

VIDAS® is found positive if test value is >=0.05

From Table 3.3 it can be noticed that for enrichment of food samples in Demi-Fraser followed by phage protein based capture and plating the same results are obtained as with VIDAS® LMO II. Detection of *L. monocytogenes* using the VIT-*Listeria* Kit after enrichment in Demi-Fraser showed significantly less positive results (2 out of 9, both samples being vegetables) compared to combined capture with plating or VIDAS® LMO II (9 out of 9 samples positive). When the enrichment broths LX and LSB were used in combination with VIT-*Listeria* Kit, the sensitivity of the method improved and *L. monocytogenes* was detected in respectively 8 and 7 samples out of 9. This indicates the importance of choosing an appropriate enrichment broth prior to application of the VIT-*Listeria* Kit. *L. monocytogenes* detection with phage protein based capture and subsequent plating on ALOA enabled detection of *L. monocytogenes* in all natural contaminated food samples (9/9) if using LX or Demi-Fraser or in all but one sample (the shrimps) if using LSB. Less typical *L. monocytogenes* cells were recovered on ALOA using LSB compared to LX or Demi-Fraser. Also more atypical flora was seen on the ALOA plates using LSB. This is probably due to the less amounts of selective compounds in this enrichment broth. In the case of combined capture and plating the prior enrichment broth seemed to be less critical to the outcome of the test result.

3.5 Discussion

Studies have been accomplished during the past years comparing different enrichment protocols or various enrichment broths (Balamurugan *et al.*, 2006; Beumer *et al.*, 1996; Grant, 2005; Lionberg *et al.*, 2003) but few of them looked in detail at the growth kinetics of the target bacterium in the broth (Duffy *et al.*, 1994a; Duffy *et al.*, 1994b; Stephens *et al.*, 1997). In this study focus was set on the determination of μ_{max} and λ for *L. monocytogenes* in various broths. To establish these growth parameters, growth curves were determined by optical density measurements (Begot *et al.*, 1996).

The first criterion to select an appropriate enrichment broth as a prerequisite for rapid detection is μ_{max}. On this basis LX, LSB and OEB can be classified as enrichment media with intermediate capacity to support growth, because their μ_{max} was significantly lower than μ_{max} of TSB and higher of Demi-Fraser (the recommended (selective) primary enrichment medium by ISO 11290-1 showing the lowest μ_{max}). LSB can be established as the best growth supporting selective enrichment medium tested because μ_{max} was significantly greater compared to the other selective enrichment media and growth was comparable to a non-selective broth.

The second criterion to select an appropriate enrichment broth as a prerequisite for rapid detection is λ. Cells that experienced the exact same stress treatment showed significant differences in λ when they were inoculated in different broths. In general selective broths induced a prolonged λ compared to non-selective broths due to the presence of selective compounds (additional stress factors), increasing time to reach detection limit of a certain method (Mellefont *et al.*, 2005; Van Houteghem *et al.*, 2008). In the present study the selective broths LX and Demi-Fraser (although not LSB) had a tendency to increase the lag phase in comparison to the non-selective TSB broths, however, not significantly. In case of exposure to oxidative stress the four selective broths failed to recover the injured cells although non-selective broth resuscitated the pathogen.

Reducing lag phase by elimination of selective agents would have benefits in recovering bacteria from food and environmental samples more rapidly. However, non-selective enrichment of foods is not readily feasible as it detracts from the ability to suppress the indigenous flora. It was shown in the present study that inclusion of selectivity in the enrichment broth is a prerequisite for reliable detection of the pathogen in naturally contaminated foods. Notwithstanding the high μ_{max} and the short lag phase, enrichment using TSB failed to detect *L. monocytogenes* in approximately 50% of the analyses performed in artificially contaminated foods with both phage protein based capture/plating or VIT-*Listeria* Kit. This finding also jeopardizes the idea of using one common non-selective shortened enrichment protocol to support the growth of a multitude of pathogenic microorganisms prior to their simultaneous detection using e.g. multiplex PCR or microarrays including several species specific target sequences. The results of λ and μ_{max} of LX and LSB and the obtained results of detection of *L. monocytogenes* in naturally contaminated samples with these broths show a correlation with regard to selectivity. Probably less selective compound are present in LSB compared to LX, which resulted in smaller λ and faster μ_{max} of LSB, but also in more difficulties for *L. monocytogenes* regarding growth (less CFU present on the isolation medium) in this broth due to the presence of accompanying flora.

From the spiked food samples it became clear that a shortened enrichment procedure (approximately 6 h) is not feasible for reliable detection of low numbers (approximately 1/g) of *L. monocytogenes*, neither for the combined phage protein based capture/plating, nor for the VIT-*Listeria* Kit. In the cases where detection occurred often less than 15 colonies were isolated on the ALOA plates or few cells were recorded in the microscopic field indicating the detection methods were performing close to their detection limit. These results were to be expected based on the results presented in Figures 3.3 and 3.4. A 24 hour prior incubation using LX broth enabled reaching of the threshold of the subsequent detection method with both phage protein based capture/plating and the VIT-*Listeria* Kit for all five food matrices (endive, deli-salad, cheese, smoked salmon, pâté). Looking at lab-flow, an "after working hours" enrichment protocol (16-18 h) is often preferred because of the ability to receive samples until late afternoon and start processing enriched samples the morning after.

To establish the suitability of an enrichment protocol, evaluation should also include naturally contaminated food in order to assess the potential for resuscitation and growth support of injured cells as present in real food samples. Therefore using naturally contaminated samples the three selective enrichment broths (Demi-Fraser, LX, LSB) were further evaluated using 17 hours incubation prior to either combined phage protein based capture and plating or VIT-*Listeria* Kit for detection. It was shown that the double-enrichment procedure (primary and secondary) as recommended for VIDAS® LMO II as well as that included in the international recognized ISO 11290:1996 method (International Standards Organization, 1996) which requires at least 48 hours can be reduced to a single-step "after working hours" enrichment procedure when one of the newly introduced optimized selective enrichment broths (LX or LSB) is used. In combination with the VIT-*Listeria* Kit, a methodology based on fluorescent *in situ* hybridization targeting 16S rRNA sequences of *L. monocytogenes* a 24 hours detection method is obtained (Stephan *et al.*, 2002; Stephan *et al.*, 2003). For some labs the need for an epifluorescence microscope to execute the VIT-*Listeria* Kit may be a drawback. The combination of phage protein based capture with plating on a chromogenic medium is a variation of the classical culture method and is a readily accessible methodology. This approach provides detection of *L. monocytogenes* as applied within 48 hours. In the present study highly specific phage proteins linked to magnetic beads were used for recognition and capture of *L. monocytogenes* (Loessner *et al.*, 2002). Enrichment combined with capture (either phage protein based capture (Hagens and Loessner, 2007; Kretzer *et al.*, 2007) or immuno-capture (Fluit *et al.*, 1993; Safarik *et al.*, 1995; Uyttendaele *et al.*, 2000) is also an approved sample preparation method prior to application of PCR. Alternatively, the captured cells in the present study may be used for DNA extraction and subsequent PCR to provide a 24 h detection method.

3.6 Conclusion

It was shown in this Chapter 3, based on the kinetics of resuscitation and the growth of *L. monocytogenes* in various enrichment broths that detection of low numbers of sub-lethally injured cells of this pathogen requires at least an overnight enrichment. A selective enrichment broth is needed to enable proliferation of *L. monocytogenes* within the indigenous bacterial flora present in foods. However, combination of an appropriate enrichment condition with advanced detection techniques may enable a 24 h detection of *L. monocytogenes*. More extensive validation studies with more food matrices and in various laboratory settings will have to confirm these findings.

Besides Gram positive *L. monocytogenes* also the Gram negative pathogens *E. coli* O157, *Campylobacter* spp. and *Salmonella* spp. are of interest in governmental food inspection, in international trade, in commercial relationships, in the food industry and in surveillance studies. In Chapter 4 enrichment broths were compared for their capacity regarding resuscitation of sub-lethally injured *E. coli* O157 cells and a similar experimental setup was used as described in Chapter 3.

Chapter 4

Comparison of enrichment conditions for rapid detection of low numbers of sub-lethally injured *Escherichia coli* O157 in food

Redrafted from:

Vicky Jasson, Andreja Rajkovic, Leen Baert, Johan Debevere and Mieke Uyttendaele

(2009)

Journal of Food Protection, 72, 1862-1868

Chapter 4 Comparison of enrichment conditions for rapid detection of low numbers of sub-lethally injured *Escherichia coli* O157 in food

4.1 *Abstract*

A comparative study of lag phase and growth rate of healthy, stressed and sub-lethally injured *Escherichia coli* O157 in ten enrichment broths was performed. The evaluation of enrichment protocols was validated by different end-point detection methods (two PCR and two combined capture/plate methods). TSB (3) provided the fastest growth (μ_{max} = 1.00 ± 0.06 h^{-1}) but failed to recover oxidative stressed *E. coli* O157. TSB (1), TSBYE, TSB supplemented with 8 mg/l novobiocin + 16 mg/l vancomycin (TSB+), BPW and BPW supplemented with 8 mg/l vancomycin (BPW+V) enabled resuscitation of *E. coli* O157 independent from pre-cultural conditions. mTSB+N, EC, ECred, TSB (3) and TSB supplemented with 8 mg/l novobiocin + 16 mg/l vancomycin + 2 mg/l rifampycin + 1 mg/l K-Telluriet + 1.5 g/l Bile salts *N°3* (TSB++) all failed recovery of *E. coli* O157 for at least one stress. The use of TSB (1), TSB+, BPW and BPW+V was compared to mTSB+N (ISO reference broth) for reliable detection of low numbers of healthy, stressed and sub-lethally injured *E. coli* O157 (approximately 1 CFU/g) from foods (sprouted seeds, fermented sausage, raw milk, raw ground beef). When low numbers of healthy cells were inoculated, BPW/BPW+V, TSB, TSB+ and mTSB+N enabled growth until detectable numbers within 6 hours enrichment at 41.5°C. Results showed that mTSB+N failed to recover freeze and "food" sub-lethally injured *E. coli* O157 until detectable numbers in contrast to BPW/BPW+V. This study highlights that using mTSB+N for recovery of *E. coli* O157 from foods may yield false negative results.

4.2 Introduction

Consumption of undercooked ground beef, fermented sausage, raw milk, dairy products, fresh produce, sprouted seeds and apple cider have all been associated with food-borne outbreaks of *Escherichia coli* O157 (Oksuz *et al.*, 2004; Rangel *et al.*, 2005). Low infectious dose (1-100 cells) (Teunis *et al.*, 2004), the ability to survive in various foods (Duffy *et al.*, 2006; Evrendilek, 2007; Muthukumarasamy and Holley, 2007) and the severity of illness (Xicohtencati-Cortes *et al.*, 2007) make *E. coli* O157 one of the most notorious food-borne pathogens (McClure and Hall, 2000; Molina *et al.*, 2003). As most of the implicated foods are highly perishable, rapid and reliable, methodology for *E. coli* O157 detection is needed. Thus, the detection procedure must provide minimal false positive result so the product is not needlessly discarded. Meanwhile, false negative result cannot be tolerated because they would allow contaminated products to be released.

Due to low numbers of *E. coli* O157 and their stressed or injured state, resuscitation and enrichment are required prior selective plating (Deisingh and Thompson, 2004). It is therefore of utmost importance to incorporate an appropriate sample preparation protocol including an enrichment step in order to avoid false negative results. The first step of a sample preparation protocol is the selection of an appropriate enrichment broth capable of resuscitating sub-lethally injured target in the presence of non-target accompanying flora. Nevertheless, suppression of non-target microorganisms by selective supplements in enrichment broth should not compromise the fast recovery of injured and physiologically deficient target microorganisms (Chapter 3). The selectivity, needed for detection, can be added in the actual detection step, which follows the enrichment step. Cells can be selectively taken from a suspension by for example antibodies attached to a solid phase or selectively detected by PCR (Gooding and Choudary, 1998; Vernozy-Rozand *et al.*, 2000).

In this study the resuscitation and enrichment efficiency of ten different enrichment broths were evaluated using lag phase (λ) and maximum specific growth rate (μ_{max}) obtained in each of these broths. The experimental setup included both healthy cells and cells stressed by freeze, "food" and oxidative stress (Chapter 2). Next, selected enrichment broths were combined with either real-time PCR or two selected magnetic separation techniques, coupled with subsequent plating on selective agar media, to assess their capacity to provide fast and specific detection of low numbers of *E. coli* O157 (approximately 1 CFU/g).

4.3 Materials and methods

4.3.1 Bacterial strains and culture conditions

E. coli O157 LFMFP 463 and LFMFP 474 were used. Culture conditions are described in detail in section 2.3.1 of Chapter 2.

4.3.2 Broths used

Ten broths were selected: **TSB** (brand 1, brand 3), **TSBYE**, **mTSB+N** [modifiedTSB (Oxoid) + 10 mg/l novobiocin (Oxoid)], **EC** (Oxoid), **ECred** [EC reduced bile salts (Oxoid)], **TSB+** [TSB (Oxoid) + 8 mg/l novobiocin (Sigma, Bornem, Belgium) + 16 mg/l vancomycin (Sigma)], **TSB++** [TSB (Oxoid) + 8 mg/l novobiocin + 16 mg/l vancomycin + 2 mg/l rifampycin (Sigma) + 1 mg/l K-Telluriet (Sigma-Aldrich, Bornem, Belgium) + 1.5 g/l Bile salts *N° 3* (Oxoid)], **BPW** [Buffered Peptone Water (BioMérieux, Marcy l'Etoile, France)] and **BPW + V** [BPW (BioMérieux) + 8 mg/l vancomycin].

4.3.3 Determination of μ_{max} and $\lambda_{control}$ in various broths

Growth at 37°C was measured by optical density (OD) measurements in 10 minutes intervals using a VERSAmax^{TM} microplate reader (Molecular Devices, Sunnyvale, CA, USA) at wavelength 380 nm. Detailed description of the protocols can be found in sections 3.3.3.and 3.3.4. of Chapter 3. Triplicate experiments were performed.

The variation in μ_{max} levels was tested by one-way ANOVA (SPSS, Inc., Chicago IL., USA). The significance of observed difference in μ_{max} among different broths was assessed by Tukey test (α=0.05) in post hoc analyses. The results of the ANOVA are indicated in Table 4.1 with Tukey based homogeneous subsets (group A, B, C and D, significant difference α=0.05).

4.3.4 Determination of λ_{stress} in various broths

A standardized procedure to evoke significant percentage of sub-lethally injured *E. coli* O157 cells was determined in Chapter 2. In brief, three stress models were included; 1) freeze stress (10 days at -21°C); 2) "food" stress (pH 4.9, 13.5% NaCl, 10 days at 4°C), and 3) oxidative stress (750 µM H_2O_2, 1h at 37°C). After stress treatment, approximately 3.5 log of *E. coli* O157 culture was inoculated in the wells of the first column of the microtiter plate as described above. The same suspensions of stressed cells were used for the different enrichment broths to eliminate minor variation between stressed cultures.

4.3.5 Microbiological analysis of food samples

Ten gram per each sample was aseptically transferred to a sterile filter stomacher bag. One tenfold dilution was made in peptone physiological saline (PPS; 8.5 g/l NaCl and 1 g/l neutralized bacteriological peptone (Oxoid)) and the sample was homogenized for 60 s by means of a stomacher Seward Laboratory blender 400 (UAC House, London, England). Subsequently, a decimal dilution series in PPS was made and enumeration on plate count agar (PCA, Oxoid) was performed to determine the total aerobic plate count. After an incubation period of 3 days at 30°C, colony forming units were counted. Similarly, coliform count was determined on violet red bile lactose agar (VRBL, Bio-Rad, Marnes-la-Coquette, France) and incubated for 24 h at 37°C, *E. coli* was enumerated on COLI ID (BioMérieux) with incubation of 24 h at 44°C. All enumerations were done by pour plate technique.

4.3.5.1 Preparation and enrichment of artificially contaminated samples.

Ten grams of each food type (fermented sausage, raw minced beef, raw soy sprouts or raw milk) were artificially contaminated with approximately 10 *E. coli* O157 LFMFP 463 cells by adding 100 µl of appropriate dilutions. To mimic real case scenarios of naturally contaminated samples not only healthy cells of *E. coli* O157 LFMFP 463 were inoculated, but also stressed and sub-lethally injured *E. coli* O157 cells. The fermented sausage was inoculated with "food" stressed cells, the raw minced beef with freeze stressed cells and the raw soy sprouts with oxidative stressed cells. For the raw milk samples healthy cells, as well as, all three types of stressed/sub-lethally injured cells were inoculated. Selected enrichment broths (BPW/BPW+V, mTSB+N, TSB+ and TSB (1)) were compared for their capacity to recover cells in combination with various *E. coli* O157 detection strategies. BPW was used for raw minced beef and BPW+V for the other food matrices. Detection was achieved by two PCR-based methods in addition to two magnetic separation-base methods, as detailed in the following section.

4.3.6 Detection methods

4.3.6.1 *E. coli* O157 Phage protein-based Capture Kit (Profos, Regensburg, Germany).

Protocol was performed as described by manufacturer. Briefly 1 ml of the enriched sample and 50 µl of phage protein coated magnetic beads were incubated at room temperature in an overhead rotor for 20 minutes. Magnetic field was applied for 5 minutes and supernatant was carefully aspirated and discarded. Samples were washed 3 times with PBST-buffer (Sigma-Aldrich). Supernatant was discarded and the particle-bacteria complex was resuspended in 100 µl of PBST-buffer before plating 50 µl on Sorbitol MacConkey Agar supplemented with cefixime tellurite selective supplement (CT-SMAC; Oxoid) and 50 µl on chromogenic O157 ID agar (BioMérieux). Colony counts were obtained after 24 h incubation at 37°C.

4.3.6.2 *E. coli* O157 Anti-O157 Dynabeads capture (Dynal, Oslo, Norway).

Protocol was performed as described by manufacturer. Briefly 1 ml of the enriched sample and 20 µl of protein coated magnetic beads were incubated at room temperature in an overhead rotor for 10 minutes. Magnetic field was applied for 3 minutes and supernatant was carefully aspirated and discarded. Samples were washed 3 times with PBST-buffer. Supernatant was discarded and the particle-bacteria complex was resuspended in 100 µl of PBST-buffer before plating 50 µl on CT-SMAC and 50 µl on chromogenic O157 ID agar. Colony counts were obtained after 24 h incubation at 37°C.

4.3.6.3 TaqMan® *E. coli* O157:H7 PCR kit (Applied Biosystems, Lennik, Belgium).

Protocol was performed as described by manufacturer. Briefly 1 ml of the enriched samples was centrifuged (Centrifuge 5415C, Eppendorf, Hamburg, Germany) (room temperature, 10 minutes, 1000 x g) and supernatant was discarded. DNA was extracted by adding 100 µl of lysis reagent (PrepMan™ Ultra, Applied Biosystems) to the pellet and heating for 10 minutes at 95-100°C. DNA extracts were diluted 1/10 by adding RNase free water (Qiagen, Hilden, Germany) and 12 µl of the obtained diluted DNA solutions was used for the amplification reaction (18 µl PCR mix). Amplification consisted of 45 cycles with 15 seconds of denaturation at 95°C and 1 minute of annealing at 60°C. Internal controls were included in each PCR reaction.

4.3.6.4 iQ-Check™ *E. coli* O157:H7 PCR kit (Bio-Rad, Marnes-la-Coquette, France).

Protocol was performed as described by manufacturer. Briefly 1 ml of the enriched samples was centrifuged (room temperature, 10 minutes, 1000 x g) and supernatant was discarded. DNA was extracted by adding 200 µl of the complete lysis reagent (beads and lysis reagent) and grind at high speed for 3 ± 1 min in a vortex disrupter at room temperature. After grinding samples were heated for 10 minutes at 95-100°C. 5 µl of the obtained DNA suspension was used for the amplification reaction (45 µl PCR mix). Amplification consisted of 50 cycles with 15 seconds of denaturation at 95°C and 30 seconds of annealing at 58°C followed by extension of the DNA by Taq polymerase for 30 seconds at 72°C. Internal controls were included in each PCR reaction.

4.4 Results

4.4.1 Determination of μ_{max}.

Table 4.1. Average μ_{max} and its standard deviation at 37°C of *E. coli* O157 LFMFP 463 in 10 different broths.

Broth	μ_{max} (h^{-1})	Stdev
TSB (1)	0.94 (A,B)	0,06
TSB (3)	1.00 (A)	0,06
TSBYE	0.91 (A,B,C)	0,09
EC	0.28 (D)	0,06
EC red	0.75 (C)	0,05
mTSB+N	0.82 (B,C)	0,06
TSB+	0.88 (A,B,C)	0,03
TSB++	0.25 (D)	0,04
BPW	0.83 (A,B,C)	0,06
BPW+V	0.78 (B,C)	0,09

Tukey based homogeneous subsets (group A, B, C and D; significant difference α=0.05).

Table 4.1 shows μ_{max} of *E. coli* O157 in function of the enrichment broths used. Significantly higher μ_{max} of *E. coli* O157 in TSB (brand 1 and 3), TSBYE, TSB+ and BPW compared to an intermediate subset of μ_{max} obtained in mTSB+N and in BPW+V and to the subset of the lowest μ_{max} obtained in EC, ECred and TSB++ is shown in Table 4.1. EC, ECred and TSB++ can be considered as less growth supporting broths.

4.4.2 Determination of $\lambda_{control}$ and λ_{stress}.

Figure 4.1 shows that EC, mTSB+N and TSB++ were not capable to resuscitate the freeze stressed cells within 24 h. Moreover, TSB (3), EC and ECred did not support recovery of oxidative stressed cells within the duration of the experiment. In addition, mTSB+N and TSB++ also inhibited growth after a “food” stress.

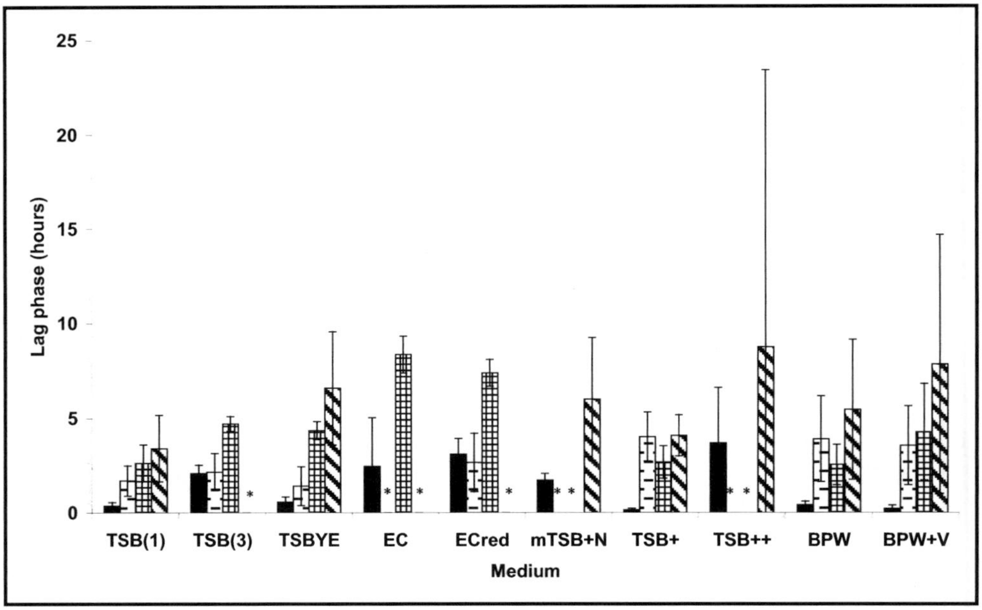

Figure 4.1. **λ of *E. coli* O157 LFMFP 463 at 37°C. Shading of bars as follows: ■ $\lambda_{control}$; ⊟ λ of freeze stressed cells; ⊞ λ of food stressed cells; ▧ λ of oxidative stressed cells. * no growth was detected. Error bars represent cumulative standard deviation.**

4.4.3 Detection of *E. coli* O157 in artificially contaminated food samples

Details of the results of the various detection strategies are presented in Table 4.2. An increase in CFU by capture combined with plating or a decrease in Ct values by real-time PCR detection can be noticed after prolonged enrichment time. In several cases a specific combination of enrichment broth and detection strategy failed to detect inoculated injured cells. The percentage of sub-lethally injured cells due to freeze, "food" and oxidative stress was determined as 99.5%, 99.9% and 76.2% (significant percentage of sub-lethal injury for the freeze and "food" stress (Chapter 2)), respectively. For the oxidative stressed cells prolongation in detection time could be noticed.

Table 4.2. Enrichment in four selected broths of healthy, stressed and sub-lethally injured *E. coli* O157 cells inoculated in four different food matrices. *E. coli* O157 cells were detected after 4, 6 and 24 h of enrichment with four different detection methods.

Matrix	Physiological state	Time (h)	*E. coli* O157 Capture				Anti-O157 dynabeads				RT-PCR AB (Ct value)				RT-PCR BioRad (Ct value)			
			BPW	mTSB+N	TSB+	TSB(1)	BPW	mTSB+N	TSB+	TSB(1)	BPW	mTSB+N	TSB+	TSB(1)	BPW	mTSB+N	TSB+	TSB(1)
Soy	Healthy cells	4	++	++	++	++	+	-	+	-	33	34	33	36	33	34	34	34
		6	+++	+++	+++	+++	+	++	++	+	29	26	27	32	28	26	25	30
		24	+++	+++	+++	+++	+++	+++	+++	+++	21	19	20	22	16	33	17	19
	Stressed cells (H_2O_2)	4	+	+	+	-	-	-	-	-	37	36	36	-	40	39	38	-
		6	++	++	++	++	+	+	+	+	33	30	30	38	30	30	37	28
		24	+++	+++	+++	+++	+++	+++	+++	+++	22	19	22	30	16	17	19	26
Raw minced beef	Healthy cells	4	+	+	+	+	+	+	+	+	35	36	35	35	37	36	38	38
		6	++	++	++	++	++	++	++	++	28	28	28	30	18	25	19	25
		24	+++	+++	+++	+++	+++	+++	+++	+++	25	20	23	27	17	15	16	19
	Sub-lethally injured cells (Freeze)	4	-	-	-	-	-	-	-	-	40	-	-	-	-	-	-	-
		6	+	-	-	+	+	-	-	+	37	-	-	-	34	-	-	-
		24	+	-	+	+	-	-	-	-	36	-	36	40	23	-	31	29
Fermented sausage	Healthy cells	4	++	++	++	++	+	+	+	-	35	34	35	35	-	-	42	40
		6	+++	+++	+++	+++	++	+++	++	++	28	27	31	30	33	33	33	31
		24	+++	+++	+++	+++	+++	+++	+++	+++	18	17	22	22	15	16	24	23
	Sub-lethally injured cells (Food)	4	+	-	-	-	-	-	-	-	-	-	-	40	-	-	-	-
		6	++	-	-	++	+	-	-	+	33	-	39	34	39	-	36	36
		24	+++	-	+	+++	+++	-	++	+	18	-	25	23	15	-	29	21
Raw milk	Healthy cells	4	++	+++	++	+++	-	-	-	-	37	35	35	33	37	33	37	32
		6	+++	+++	+++	+++	++	+++	+++	+++	28	28	27	25	25	26	25	25
		24	+++	+++	+++	+++	+++	+++	+++	+++	20	20	20	23	14	Inh.	15	Inh.
	Stressed cells (H_2O_2)	4	-	-	++	-	-	-	-	-	-	-	36	-	-	43	39	-
		6	++	+++	+++	+++	+	-	+	-	35	34	31	31	33	32	30	30
		24	+++	+++	+++	+++	+++	+	+++	+	26	24	25	29	21	22	19	Inh.
	Sub-lethally injured cells (Freeze)	4	-	-	-	-	-	-	-	-	-	-	-	-	-	-	-	-
		6	-	-	-	-	-	-	-	-	-	-	-	-	-	-	-	-
		24	-	-	-	-	-	-	++	-	-	-	-	-	-	Inh.	29	-
	Sub-lethally injured cells (Food)	4	-	-	+	-	-	-	-	-	-	-	39	-	-	-	-	38
		6	+++	-	+++	+++	+	-	+	++	31	-	33	30	29	-	32	28
		24	+++	-	+++	+++	+++	-	-	+	24	-	27	27	18	-	22	Inh.

BPW*: BPW was used for raw minced beef and BPW+V was used for soy, fermented sausage and raw milk

For MS: -: absence; +: 1 ≤ CFU/plate ≤ 14; ++: 15 ≤ CFU/plate ≤ 150; +++: ≥ 151 CFU/plate

For RT-PCR (real-time PCR): -: no signal; Inh.**: Inhibition of PCR reaction

The type of food (and associated variation in competitive flora) did not seem to influence the recovery of *E. coli* O157. The total aerobic count for soy, minced beef, fermented sausage and raw milk were respectively 7.5, 4.9, 8.2 and 3.2 log CFU/g and respectively 7 and 3.3 log CFU/g coliforms were found in soy and raw minced beef (for the fermented sausage and raw milk <10 CFU/g coliforms). None of the food samples showed detectable numbers of *E. coli* (<10 CFU/g) prior to inoculation.

In order to visualize the effect of sub-lethally injured cells on the performance of enrichment (and thus detection) strategies, results of detection of "food" and freeze stress cells (Figure 4.2B) and the parallel results of the healthy cells (Figure 4.2A) were analyzed in terms of sensitivity rate. The sensitivity rate has been defined as the ratio of the number of positive samples found and the number of inoculated samples. A sensitivity rate of 1 means that in all cases when food samples were inoculated the combination of enrichment broth, enrichment time and detection method resulted in detection of *E. coli* O157 cells.

When the sensitivity rate was smaller than 1 but bigger then 0 at least for one food sample a certain combination of enrichment/detection protocol resulted in no detection of *E. coli* O157 cells although the food sample was inoculated. A sensitivity of 0 revealed that the food samples were inoculated but no detection of target cells could be achieved.

Figure 4.2A shows that different food samples inoculated with healthy cells of *E. coli* O157 were all found positive (sensitivity rate = 1) already after 6 h enrichment independently from enrichment broth and detection strategy. A sensitivity rate of 1 was already obtained after 4 h enrichment (independently from the selected broth) if implores detection strategy was either *E. coli* O157 phage protein-based Capture Kit (Profos) or TaqMan® real-time PCR (Applied Biosystems).

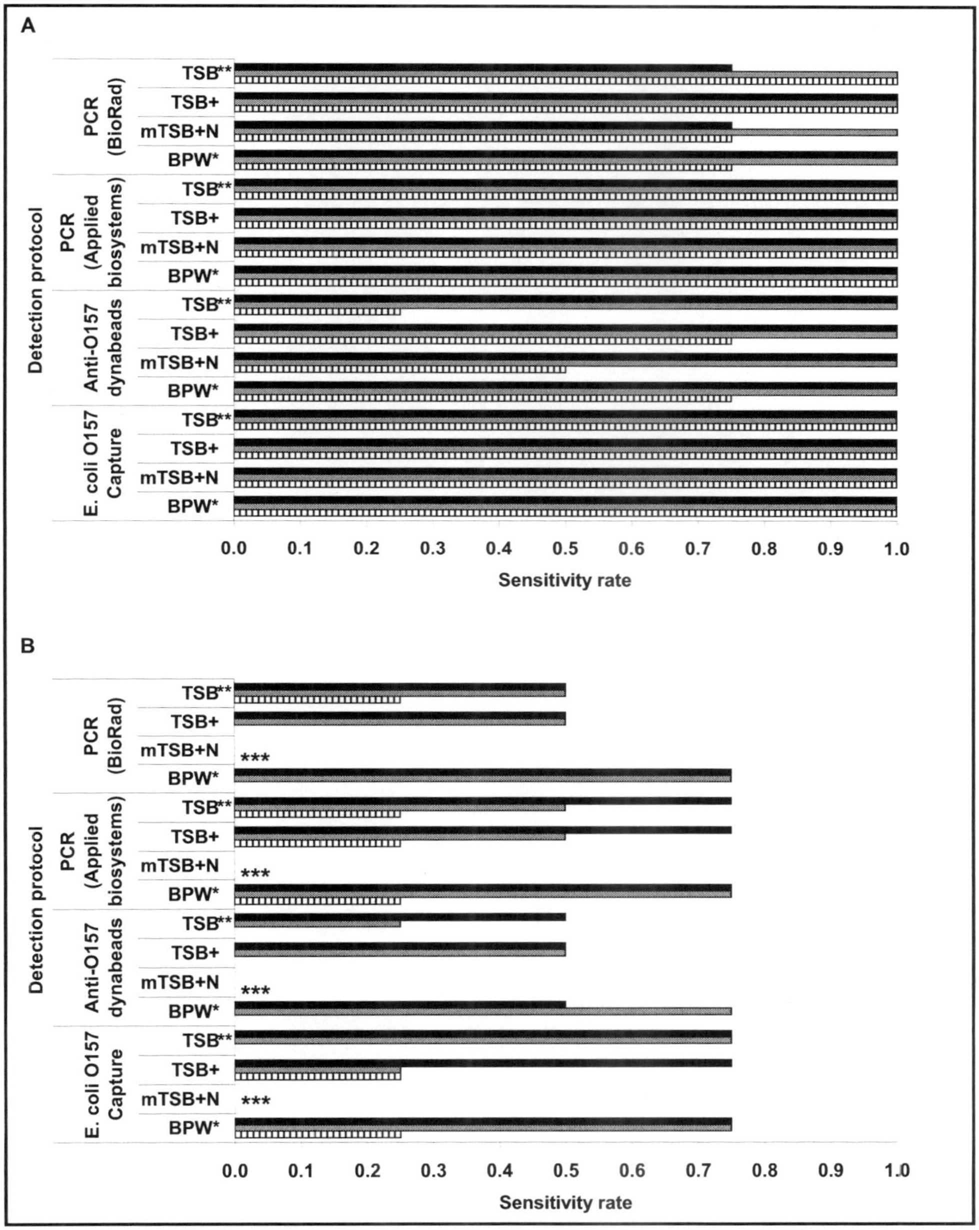

Figure 4.2. Sensitivity rate of the detection protocol (combination enrichment broth and detection method) when different food samples were inoculated with approximately 10 healthy cells (A) or sub-lethally injured cells (B)("food" and freeze stressed cells). A sensitivity rate of 1 implies that all inoculated food samples were found positive. Bars represent sensitivity rate of the detection protocol when food samples were incubated at 37°C for 4 h (▥), 6 h (▒) or 24 h (■). BPW*: BPW was used for raw minced beef and BPW+V was used for soy, fermented sausage and raw milk. TSB**: TSB (1). ***: no detection of *E. coli* O157 cells in food samples.

For sub-lethally injured *E. coli* O157 cells (> 99% sub-lethal injury established), enrichment in BPW/BPW+V enabled the maximum recovery after 6 h enrichment (Figure 4.2) with the various detection methods, although a sensitivity rate of 1 was not reached. Even 6 h of incubation in BPW/BPW+V for many detection strategies was at detection limit indicated by only few colonies isolated by capture on plate or the high Ct values. In some cases injured cells were detected after 4 h enrichment with a specific detection strategy e.g. freeze stressed cells inoculated in raw minced meat and food stressed inoculated in fermented sausage both in the case of real time PCR detection. However, in both cases a Ct value of 40 was obtained. mTSB+N did not enable the resuscitation of freeze stressed and "food" stressed sub-lethally injured *E. coli* O157 cells.

4.5 Discussion

Detection of sub-lethally injured cells, especially pathogens characterized with low infectious dose, such as *E. coli* O157 is critical (Wu, 2008). The selectivity of the enrichment broth is therefore a major point of attention as selective compounds such as bile salts and antibiotics could inhibit the recovery of injured cells (Blackburn and McCarthy, 2000). The first objective of this study was to determine the best enrichment broth for *E. coli* O157. Based on the combination on short λ, a fast μ_{max} and resuscitation capacity that was independent from pre-cultural conditions, BPW/BPW+V, TSB (1) and TSB+ were selected for further evaluation. In the present study, significantly higher μ_{max} of *E. coli* O157 was determined in the non-selective media compared to the classical selective enrichment media (Table 4.1). However non-selective broths for *E. coli* O157 cannot generally be recommended for food samples with high level of background flora, as growth of such flora can outcompete target cells (Besse *et al.*, 2005). This is the reason why in recommended ISO methods for *E. coli* O157 ISO 16654:2001 (International Standards Organization, 2001a) the selective broth mTSB+N is recommended for enrichment purposes.

The second objective of this study was to evaluate the use of non-selective media as enrichment broths in the detection of *E. coli* O157 in foods with indigenous microflora. For this purpose healthy, stressed, sub-lethally injured cells were inoculated in different food samples. Different detection methods were included in the experimental setup to account for a complete strategy with different detection methods. The combination of detection technologies and formats supports the selection of an appropriate enrichment broth. In the present study, it was evident that mTSB+N consistently failed to allow resuscitation and enrichment of sub-lethally injured *E. coli* O157 cells. Similar false negatives were reported elsewhere (Blackburn and McCarthy, 2000; Hepburn *et al.*, 2002). Based on the data obtained preference is given to a non-selective medium such as BPW for enrichment of *E. coli* O157 from foods. This is further supported by comparison of μ_{max} of *E. coli* O157 ($0.94 \pm 0.06\ h^{-1}$) in TSB and μ_{max} of *L. monocytogenes* in TSB (0.39 ± 0.05) (Chapter 3) it can be noticed that the μ_{max} of *E. coli* O157 is higher than the μ_{max} of *L. monocytogenes.* This implies that *E. coli* O157 is a fast grower comparable to *Salmonella* (ISO 6579:2005)(International Standards Organization, 2005b) enabling the option of a non selective medium (BPW or Brain Heart Infusion broth) or medium with limited antibiotics (BPW+V) for enrichment without running a risk of being overgrown by the competitive flora. This has been confirmed in the present study by the experiments with the various food types.

This study emphasizes the shift in performance of detection strategies (and especially the choice of the prior enrichment broth for the execution of the detection method) if using healthy cells versus sub-lethally injured cells. It is therefore highly recommendable to use naturally contaminated samples or samples inoculated with injured cells when designing and evaluating detection strategies protocols (Chapter 2). Due to low prevalence of *E. coli* O157, finding naturally contaminated food samples is not that evident (Carney *et al.*, 2006; Matthews *et al.*, 2006). In order to compensate for low prevalence of *E. coli* O157 in foods and still mimic natural contamination the present study was based on stressed and sub-lethally injured inoculum. The incorporation of cultures with stress exposure history differentiated this present study from elsewhere published (Gooding and Choudary, 1998; Grant, 2005; Grant, 2008; Hussein *et al.*, 2008).

The selectivity needed for the detection strategy can be added in the actual detection method (magnetic separation combined with plating, PCR, ELISA) which follows the enrichment step. Immunomagnetic separation using O157-specific-antibody-coated beads followed by bacteriological cultures has been shown to improve the detection of *E. coli* O157 in foods (Okrend *et al.*, 1992; Wright *et al.*, 1994) and is included as part of the ISO standard method (International Standards Organization, 2001a). In this case the sensitivity and the specificity of the detection technique are in the first place based on the choice of the antibody early during development of a certain kit.

In the present study the phage protein based capture supported detection of *E. coli* O157 (both healthy and injured cells) overall faster (i.e. occasionally already after 4 h enrichment time) than the antibody-based capture. This can be explained by the higher affinity (and thus sensitivity) of bacteriophage proteins. Phages proteins (from bacteriophages) are more specifically attached to bacteria compared to antibodies (Kretzer *et al.*, 2007).

However the inclusion of capture step takes its toll in the total length of the detection procedure as it requires still a 24 h incubation of the captured cells on the selective (chromogenic) agar media. A solution to this is to use the captured cells for DNA extraction and PCR detection to obtain a faster result. However, from the present study it has been shown that concentration of cells by capture is not a necessary intermediate step for successful PCR detection. It was demonstrated that one ml of 6 h in BPW enriched culture can be directly processed for DNA extraction and PCR providing result within 8 h. This has been shown to be a rapid and reliable detection strategy for recovery of low numbers (of also injured) *E. coli* O157 cells in food. Further experiments should be performed to determine statistically the best detection strategy for fast and reliable detection of *E. coli* O157 from different food samples after a non or less selective enrichment broth is used. These further experiments should be performed with a wider range of foods and different replicates, preferably including an interlaboratory trial.

4.6 Conclusion

In the present chapter, significantly higher μ_{max} of *E. coli* O157 was determined in the non-selective media compared to the classical selective enrichment media. It was demonstrated that mTSB+N, the recommended enrichment broth in ISO 16654:2001 consistently failed to allow resuscitation and enrichment of sub-lethally injured *E. coli* O157 cells. Based on the data obtained preference is given to a non-selective medium such as BPW for enrichment of *E. coli* O157 from foods.

Further experiments should be performed to determine statistically the best detection strategy for fast and reliable detection of *E. coli* O157 from different food samples after a non or less selective enrichment broth is used. These further experiments should be performed with a wider range of foods and different replicates, preferably including an interlaboratory trial.

In Chapter 5 Bolton broth, the recommended enrichment broth in ISO 10272:2006 for the detection of *Campylobacter* spp. from food, feed and environmental samples was tested for it performance regarding detection of freeze stressed and healthy *Campylobacter* spp.

Chapter 5

Characterization of *Escherichia coli* from raw poultry in Belgium and impact on the detection of *Campylobacter jejuni* using Bolton broth

Redrafted after:

Vicky Jasson, Imca Sampers, Nadine Botteldoorn, Francisco López-Gálvez, Leen Baert, Sarah Denayer, Andreja Rajkovic, Ihab Habib, Lieven De Zutter, Johan Debevere and Mieke Uyttendaele

(2009)

International Journal of Food Microbiology, 135, 248-253

Chapter 5 Characterization of *Escherichia coli* from raw poultry in Belgium and impact on the detection of *Campylobacter jejuni* using Bolton broth

5.1 Abstract

A comparative study examining Bolton broth and Preston broth for enrichment and reliable detection of *Campylobacter jejuni* (both healthy and freeze stressed cells) was performed. Tested as pure cultures, Bolton broth enabled faster resuscitation and growth of *C. jejuni* compared to Preston broth. When *C. jejuni* was co-incubated with extended-spectrum-beta-lactamase (ESBL) producing *Escherichia coli* isolated from Belgian poultry meat preparations, the latter dominated in the Bolton enrichment broth and crowded the mCCDA plates. This resulted in the inability to recover *C. jejuni* by ISO 10272:2006 standard method. Preston broth did not support the growth of the ESBL *E. coli* isolates, but showed longer time to detection of *C. jejuni* compared to Bolton broth. The use of the same antibiotic (sodium cefoperazone) in Bolton broth and in mCCDA plates may explain the problems encountered for detection of *C. jejuni*, as high numbers of ESBL *E. coli* present after enrichment in Bolton broth, also caused overgrowth and masked the few *C. jejuni* colonies present on the mCCDA plates. The use of *Campylobacter* spp. specific real-time PCR circumvented these problems and enabled rapid detection of the pathogen after 24 h enrichment in both Bolton and Preston broth, for both healthy and freeze stressed cells.

5.2 Introduction

Campylobacter spp. are the major cause of food-borne bacterial gastroenteritis in Europe (EFSA, 2009). Poultry has often been associated with campylobacteriosis (Humphrey *et al.*, 2007; Vellinga and Van Loock, 2002). In most of the cases only low numbers are present in poultry meat (especially skinless poultry meat or meat preparations) (Callicott *et al.*, 2008; Habib *et al.*, 2008a; Sampers *et al.*, 2008). Also due to food preservation techniques such as freezing, the bacteria might be stressed or sub-lethally injured and may fail to grow on selective media after direct plating (Chapter 2). Therefore a qualitative method based on enrichment of *Campylobacter* spp. is often used when establishing monitoring programs in the poultry chain. During enrichment, the target bacteria are allowed to resuscitate and enough numbers are reached for subsequent isolation by plating or alternative detection methods (e.g. ELISA and PCR).

The ISO 10272:2006 method for the detection of *Campylobacter* spp. in food (International Standards Organization, 2006b), recommends the use of Bolton broth. This broth contains limited amounts of selective compounds as these selective compounds may inhibit the fast recovery of *Campylobacter* spp. (Baylis *et al.*, 2000). In a former version of ISO 10272:1995 (International Standards Organization, 1995) the more selective Preston broth was recommended as enrichment medium for the detection of *Campylobacter* spp. and is still used in some countries (Baylis *et al.*, 2000b; Paulsen *et al.*, 2005).

Various types of raw poultry meat are available at retail including whole carcasses and poultry meat cuts (fresh or frozen), as well as chicken fillet and more recently chicken meat preparation. The latter become increasingly popular with the Belgian consumer. Chicken meat preparations can be defined as raw minced chicken meat (with or without skin) that is produced in the form of hamburgers, sausages, minced meat etc., but also include marinated or seasoned wings or filets intended to be eaten. These hamburgers or sausages can be compared with the classical ones only the beef or pork is replaced by chicken meat (Sampers *et al.*, 2008; Uyttendaele *et al.*, 2006).

Habib *et al.* (2008), in a Belgium-wide survey of *Campylobacter* spp. contamination in these chicken meat preparations, found that 41.0 % of these food samples was *Campylobacter* spp. positive by direct plating compared to only 24.2 % by enrichment. They noticed that samples with lower *Campylobacter* spp. concentrations had lower likelihood of giving a positive result by selective enrichment compared to direct plating. Similar results were obtained by Musgrove *et al.* (2001), however testing was performed using different types of food and environmental matrices.

The objective of the present study was to provide insight in discrepancies found between detection of *Campylobacte*r spp. with enrichment protocol of ISO 10272:2006 (International Standards Organization, 2006b) and with direct enumerations. The study was based on false negative results obtained by enrichment in at least 20 % of the tested samples where *Campylobacter* spp. was present at levels greater than 10 CFU/g. Atypical colonies found on mCCDA plates after enrichment of Belgian chicken meat preparation and chicken carcasses were studied in detail. Some preliminary results indicated difficulties in the detection of *Campylobacter* spp. from Belgian chicken meat when mCCDA plates were used due to competition of antibiotic resistant *Escherichia coli*. Real-time PCR was included in this study as an alternative detection method.

5.3 Materials and methods

5.3.1 *Campylobacter* strains and culture conditions

Two *Campylobacter jejuni* strains were used: LFMFP 595 = Belgian Co-ordinated Collections of Microorganisms LMG 8841, and LFMFP 116 isolated from Belgian chicken meat preparations (Table 5.1). Culture conditions for the different strains are described in detail in section 2.3.1 of Chapter 2.

Table 5.1. Used strains and their origin

Microorganism	*LFMFP No	**LMG No	Origin
C. jejuni	595	8841	Bovine
	116		Chicken meat preparation
E. coli	134		Carcass
	135		Carcass
	136		Carcass
	137		Carcass
	138		Carcass
	139		Carcass
	140		Skin
	146		Drumstick (no bone no skin)
	162		Drumstick (no bone no skin)
	181		Minced chicken meat
	199		Minced chicken meat
	202		Minced chicken meat
	237		Chicken sausage
	246		Chicken sausage
	250		Chicken sausage
	252		Chicken sausage
	255		Skin
	256		Drumstick (no bone no skin)
	263		Drumstick (no bone no skin)
	270		Hands personnel
	271		Skin
	272		Skin
	273		Skin
	274		Skin
	275		Skin
	276		Skin
	277		Skin
	278		Skin
	279		Skin
	168	8063	Faeces

***LFMFP: Laboratory of Food Microbiology and Food Preservation**

****LMG: Laboratory of Microbiology Ghent**

5.3.2 *E. coli* strains and their characterization

In total twenty-nine atypical colonies, which crowded mCCDA plates, streaked from prior 24 h Bolton enrichment of poultry carcasses, chicken meat preparations, chicken skin or drumsticks were randomly picked. These repeatedly found atypical colonies were all confirmed to be *E. coli* strains by showing typical colonies on chromogenic Coli ID medium (BioMérieux, Marcy l'Etoile, France) and identified as *E. coli* by Crystal ID (Becton Dickinson, Erembodegem, Belgium). Culture conditions for the different strains are described in detail in section 3.2.1 of Chapter 2.

5.3.3 Enrichment of *C. jejuni* in the presence of Belgian chicken meat *E. coli* isolates: co-culture experiment

To assess the capacity of Bolton and Preston broth to provide fast and specific detection of *C. jejuni* in the presence of other flora, a co-culture experiment with a cocktail of three *E. coli* isolates (LFMFP 135, LFMFP 137 and LFMFP 139) from Belgian chicken meat was performed. Bolton broth containing the Bolton antibiotic supplement and 5% lysed horse blood and Preston broth i.e. Nutrient broth (Oxoid) containing the Preston *Campylobacter* selective supplement (Oxoid) and 5% lysed horse blood were evaluated for resuscitation and enrichment purposes. Bolton and Preston broth were prepared according to manufacturer recommendations and kept for a maximum of 4 weeks at 4°C before use. Different combinations of initial concentrations of *C. jejuni* (C) (LFMFP 595 or 116) in combination with approximately 1000 CFU/ml of the 3-strain *E. coli* cocktail (E) were used to inoculate (300 µl) Bolton and Preston broth (30 ml) to establish ratios of *C. jejuni* versus *E. coli* of 1:100 (C<E), 1:1 (C=E) and 10:1 (C>E). For all experiments as control a pure culture of *C. jejuni* (C) was inoculated in Bolton or Preston broth. A first duplicate set of inoculated broths holding an inoculum of healthy cells of *C. jejuni* and *E. coli* in defined ratios was incubated under micro-aerobic conditions at 42°C and samples were analyzed for the presence of *E. coli* [spread plated on Plate Count Agar (PCA, Oxoid), aerobically incubated for 24 h at 37°C] and *C. jejuni* [(spread plated on mCCDA, incubated micro-aerobically for 24 and 48 h at 42°C] after 0, 4, 6, 8, 24 and 48 h of enrichment. In addition samples were taken for detection of *C. jejuni* by a *Campylobacter* spp. specific real-time PCR after 0, 8, 24 and 48 h of enrichment.

Secondly the recovery of *C. jejuni* in co-culture with *E. coli* after freezing was determined. The same cultures used for the inoculations were kept in the freezer at -21°C for 7 days to induce injury of *C. jejuni* and *E. coli* cells. Next, cultures were defrosted at 4°C for 1 h and used to inoculate Bolton and Preston broth maintaining the same ratios by using the same volumes and dilutions as mentioned above for the healthy cells. Again duplicate set of tubes were inoculated. Tubes were incubated under micro-aerobic conditions at 42°C and samples were analyzed for the presence of *E. coli* and *C. jejuni* after 0, 4, 6, 8, 24 and 48 hours of enrichment by means of enumeration as mentioned above and in addition by *Campylobacter* spp. specific real-time PCR.

5.3.4 *Campylobacter spp.* specific real-time PCR detection

For DNA extraction PrepMan Ultra Sample Preparation Reagent (Applied Biosystems, Lennik, Belgium) was used. Briefly, 1 ml of the enrichment broth after 0, 8, 24 and 48 h of incubation was centrifuged (Centrifuge 5415C, Eppendorf, Hamburg, Germany) (room temperature, 10 min, 1000 x g) and supernatant was discarded. 100 µl of PrepMan Ultra Sample Preparation Reagent was added and samples were heated for 10 min at 95°C. Samples were centrifuged again (room temperature, 3 min, 1000 x g) and 10 µl of DNA extract (supernatant) was added to 90 µl of nuclease free water (Qiagen, Hilden, Germany). The real-time PCR assay was carried out in a 25 µl volume and contained 1 x Taqman master mix (Applied Biosystems), primers and probe (16S rRNA gene sequence) described by Lund *et al.* (2004) and previously applied by Botteldoorn *et al.* (2008) (final concentration of 600 nmol/l of each primer and 200 nmol/l of the probe (Eurogentec, Seraing, Belgium) and 5 µl of the template DNA. Real-time PCR was performed on the ABI 7300 sequence detection system using the FSHV 1.0 application software (Applied Biosystems) with the following amplification program: an initial activation step of the enzyme at 95°C for 10 min followed by 40 cycles of 95°C for 15 s and 1 min annealing and elongation at 60°C. Real-time PCR of *C. jejuni* was executed merely for detection of *C. jejuni* as a qualitative assay. Nevertheless cycle threshold (Ct) values of the PCR detection were monitored to see if a trend of continuous and significant decreasing Ct values occurred after 24-48 h incubation, which might indicate, in a semi-quantitative manner, that the initial numbers of *C. jejuni* had increased during enrichment.

5.3.5 Characterization of the antibiotic resistance profile of *E. coli* strains isolated from Belgian chicken meat preparations

The *E. coli* strains used in the co-culture assay (LFMFP 135, 137 and 139) and one BCCM-UGent culture collection strain (faecal origin) *E. coli* LFMFP 168 (Table 5.1) were tested for their resistance towards the different antibiotics present in Bolton (20.0 µg/ml sodium cefoperazone, 20.0 µg/ml trimethoprim and 20.0 µg/ml vancomycin) according to CLSI M100-S17 microtiter plate method (Clinical and Laboratory Standards Institute, 2007). Further the antibiotic susceptibility of all 29 randomly picked *E. coli* strains, isolated from mCCDA-plates (Table 5.1), was analyzed by the commercially available broth dilution method, Sensititre® (Trek Diagnostic Systems, West Sussex, England) according to the manufacturers' protocol. The tested antibiotics were sulphametoxazole (Sul), gentamicin (Gm), ampicillin (Amp), cefotaxime (Fot), ceftazidime (Taz), tetracyclin (Tet), streptomycin (Str), trimethoprim (Trmp), chloramphenicol (Chl), florphenicol (Flor), kanamycin (Kan), nalidixic acid (Nal), ciprofloxacin (Cip) and colistin (Col). To distinguish between resistant and sensitive the breakpoints proposed by eucast (EUCAST, 2009) were used. Seventeen strains showing different resistance profiles were selected for further molecular analysis in order to determine their extended-spectrum-beta-lactamase (ESBL) properties. This was done using PCR amplification of the beta-lactamases of the CTX, SHV and TEM families, and subsequent sequencing of the amplicons as described previously to detect small variances (Bertrand *et al.*, 2006).

5.4 Results

For the comparison of Bolton versus Preston broth with regard to the resuscitation and enrichment capacities of *C. jejuni*, either as a pure culture or as a co-culture with a mixture of Belgian poultry meat specific *E. coli* isolates, growth kinetics were monitored, both for healthy and freeze stressed cells. Figure 5.1A shows the growth kinetics of *C. jejuni* LFMFP 116 during enrichment in Bolton broth. When only *C. jejuni* was present significant growth in Bolton broth was detected on mCCDA plates after 4 h of incubation. For the growth of *C. jejuni* in combination with *E. coli* no growth of *C. jejuni* was detected after 4 h. Problems regarding overgrowth of the mCCDA plates by competing *E. coli* occurred after 6 h of enrichment. Either no *C. jejuni* colonies were detected anymore on the mCCDA plates or they were extremely difficult to count. When detection of *C. jejuni* LFMFP 116 with real-time PCR was considered, a decrease in Ct values compared to time 0 (6-9 Ct values after 24-48 h incubation) was noticed along with the incubation time in Bolton broth, indicating an increase in *C. jejuni* DNA, indicating growth and thus multiplication of *C. jejuni* in the co-culture.

Figure 5.1C shows the growth kinetics of *C. jejuni* LFMFP 116 during enrichment in Preston broth. *C. jejuni* present in a lower concentration than *E. coli* (C<E) was also included. It can be noticed that Preston broth supported the growth of *C. jejuni* LFMFP 116, but only after 8 h of enrichment. Also in Preston broth a decrease in Ct values was noticed along with the incubation time, indicating increasing level of *C. jejuni* DNA and thus multiplication of *C. jejuni*.

Figure 5.1B shows the growth kinetics of freeze stressed *C. jejuni* LFMFP 116 during enrichment in Bolton broth. A decrease in the numbers of *C. jejuni* LFMFP 116 of approximately 1.2 log CFU/ml after freezing was noted on mCCDA plates compared to the corresponding inoculum of healthy cells (Figure 5.1A). In the case of simultaneous inoculation with *E. coli*, it can be noticed that no *C. jejuni* cells were recovered on the mCCDA plates after 6 h of incubation not even when *C. jejuni* were present in higher concentration compared to *E. coli*. Nevertheless, a decrease in Ct value was noted along with the incubation time of 48 h indicating growth of *C. jejuni* cells.

Figure 5.1D shows the growth kinetics of *C. jejuni* LFMFP 116 during enrichment in Preston broth for an inoculum of freeze stressed cells. Figure 5.1D and Figure 5.1C can be used to compare the behaviour of healthy versus freeze stressed cells in Preston broth. If a 24 h enrichment time was respected, in all cases *C. jejuni* cells were recovered on the mCCDA plates as well as by real-time PCR. A 48 h enrichment period provided even higher numbers of *C. jejuni* in Preston broth (and lower Ct values) compared to 24 h incubation.

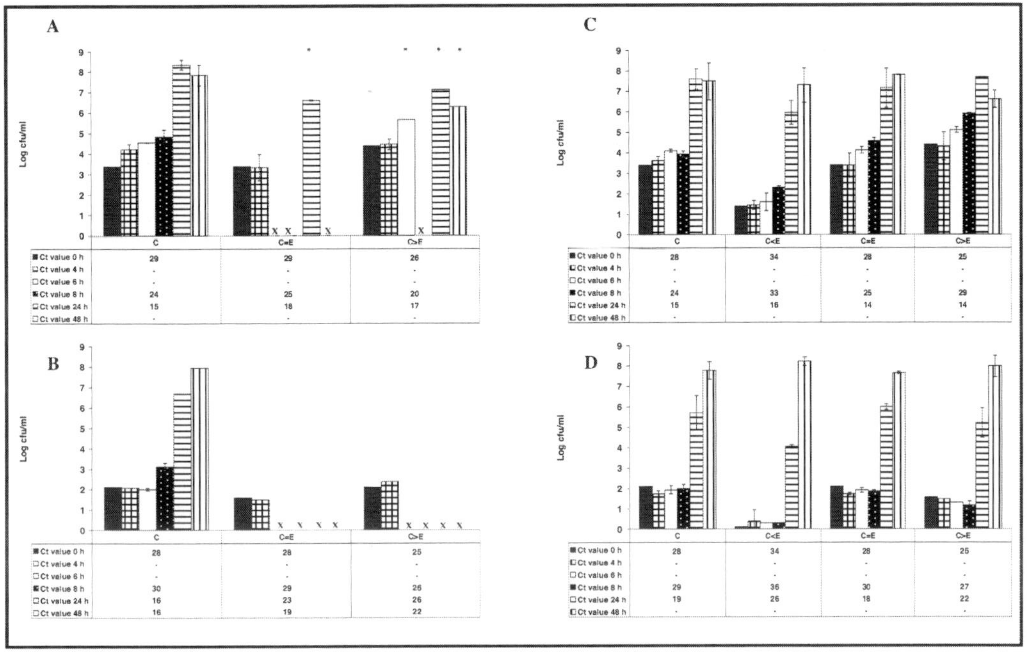

Figure 5.1. Detection of *C. jejuni* LFMFP 116 in co-culture assay with a mixture of *E. coli* strains by means of selective plating on mCCDA and real-time PCR after enrichment in Bolton broth before (5.1A) and after freezing (5.1B) or Preston broth before (5.1C) and after freezing (5.1D). *C. jejuni* was inoculated as a pure culture (C) or ratios of *C. jejuni* versus *E. coli* of were inoculated: 1:100 (C<E), 1:1 (C=E) and 10:1 (C>E). Samples were analyzed for the presence of *C. jejuni* after 0 ■, 4 ▦, 6 □, 8 ■, 24 ▤ and 48 h ▥ of enrichment at 42°C under micro-aerobic conditions. Log CFU/ml represents data of duplicate experiments. X: no *C. jejuni* cells were detected on mCCDA plates; *: log CFU/ml counts are only an indication of *C. jejuni* present (mCCDA plates were crowded with atypical colonies); -: no data obtained.

Enumeration data of *E. coli* obtained after various incubation times of enrichment in Preston broth showed the inability for these *E. coli* isolates to grow in the selective Preston broth (data not shown). Already after 4 h of enrichment in Preston broth, no *E. coli* could be recovered (i.e. < 1 CFU/ml). On the contrary after 6 and 8 h of enrichment in Bolton broth respectively, 6 and 7 log CFU/ml *E. coli* were present for both healthy and freeze stressed cells.

The whole experimental setup of co-cultures was repeated with the same specific *E. coli* 3-strains cocktail and another *C. jejuni* strain and similar results were obtained (data not shown).

Table 5.2. Minimum inhibitory concentration of different isolated *E. coli* strains.

Antibiotics	MIC* (µg/ml)			
	LFMFP** 168	LFMFP 135	LFMFP 137	LFMFP 139
Trimethoprim	80	80	80	80
Sodium cefoperazone	<0,078	>80	>80	>80
Vancomycin	>80	>80	>80	>80

*MIC: Minimum inhibitory concentration

** LFMFP: Laboratory of Food Microbiology and Food Preservation

The antibiotic resistance profile of the *E. coli* strains isolated from Belgian chicken meat preparation was determined. *E. coli* LFMFP 168, a reference strain, did not show growth in Bolton broth in contrast to the Belgian chicken meat isolated *E. coli* LFMFP 135, 137 and 139 strains (data not shown). Table 5.2 shows that the MIC of sodium cefoperazone for *E. coli* LFMFP 168 is 0.078 µg/ml which is less compared to the concentration of sodium cefoperazone present in Bolton broth (20.0 µg/ml sodium cefoperazone, 20.0 µg/ml trimethoprim and 20.0 µg/ml vancomycin). This indicates that sodium cefoperazone prevents the growth of *E. coli* LFMFP 168 in Bolton broth.

Table 5.3. *E. coli* strains selected for antibiotic susceptibility analysis using the Sensititre® method and molecular analysis of the ESBL-type. Break points to distinguish between 'Resistance' and 'Sensitivity' were defined as stated by EUCAST.

LFMFP No*	Multiresistance profile	Profile No	ESBL- type		
			SHV	TEM	CTX
134	SulAmpFotTazTetStrTrmpKanNalCip	1	-	TEM-1	+
135	SulAmpFotTazTetStrTrmpKanNalCip	1	-	TEM-1	+
136	SulAmpFotTazTrmpNalCip	2	-	-	CTX-M-1
137	SulAmpFotTazTrmpNalCip	2	-	-	CTX-M-1
138	SulAmpFotTazTetStrTrmpKanNalCip	1	-	TEM-1	+
139	SulAmpFotTazTetStrTrmpKanNalCip	1	-	TEM-1	+
140	SulAmpFotTazStrTrmpChlKanNalCip	3	-	-	CTX-M-1
146	SulAmpFotTazTetStrTrmp	4	ND	ND	ND
162	SulAmpFotTazTetTrmpChlNalCip	5	-	-	CTX-M-56
181	AmpFotTazTetChl	6	-	-	CTX-M-14
199	SulAmpFotTazStrTrmp	7	-	-	+
202	SulAmpFotTazTetTrmp	8	-	-	CTX-M-1
237	SulAmpFotTazTetStrTrmp	4	ND	ND	ND
246	SulAmpFotTazTetStrNalCip	9	-	-	+
250	AmpFotTazTetTrmpNalCip	10	-	-	CTX-M-1
252	AmpFotTazTetTrmpNalCip	10	-	-	CTX-M-1
255	AmpFotTazTet	11	+	-	CTX-M-1
256	SulAmpFotTazTetStrTrmpNalCip	12	-	-	+
263	SulAmpFotTazTetStrTrmpNalCip	12	-	-	+
270	SulAmpFotTazTetTrmpNalCip	13	-	TEM-1	CTX-M-56
271	SulAmpFotTazTetTrmpChlNalCip	5	-	-	CTX-M-56
272	SulAmpFotTazTetTrmpChlNalCip	5	-	-	CTX-M-56
273	SulAmpFotTazTetTrmpChlNalCip	5	-	-	CTX-M-56
274	SulAmpFotTazTetStrTrmpCip	14	-	TEM-1	CTX-M-56
275	SulAmpFotTazTetStrTrmpChlKanNalCip	15	-	TEM-1	CTX-M-35
276	SulAmpFotTazTetStrTrmpChlKanNalCip	15	-	TEM-1	CTX-M-35
277	SulAmpFotTazTetStrTrmpChlNalCip	16	-	TEM-1	-
278	SulAmpFotTazStrTrmpChlNalCip	17	-	TEM-1	CTX-M-15
279	SulAmpFotTazNalCip	18	+	-	CTX-M-1

*** LFMFP: Laboratory of Food Microbiology and Food Preservation**

ND: not determined

-: *bla* gene not present

+: *bla* gene present

Trimethoprim and vancomycin do not inhibit the growth of *E. coli*, but are added to the medium to inhibit the growth of *Proteus* spp. and Gram positive bacteria respectively. It can be noticed from Table 5.2 that growth of *E. coli* LFMFP 135, 137 and 139 is possible in Bolton broth as these strains show resistance towards a concentration of 80 µg/ml of sodium cefoperazone. Table 5.3 shows that none of the 29 selected *E. coli* strains tested for the different antibiotic were resistant to Gm, Flor or Col, while all strains were resistant to Amp, Fot and Taz. High levels of resistance were also found for Nal (76% resistant), Sul (86% resistant), Tet (79% resistant), Trmp (86% resistant) and Cip (79% resistant). All strains were multi-resistant (resistance to at least 4 antibiotics), with 6 strains multi-resistant to 10 antibiotics and 2 strains which were even resistant to 11 of the antibiotics tested (Table 5.3). Within these 29 strains, 18 different resistance profiles were found. Because of the high level of resistance against ampicillin and the cephalosporin's of the 3rd generation, ESBL was confirmed. Molecular methods used to study the beta lactam type of a representative strain for each profile, demonstrated the presence of the bla_{CTX}-gene in most of the strains (16), with 5 of these strains also having the bla_{TEM}-gene. In 2 strains the bla_{SHV}-gene was also present. Strain LFMPF 277 only possessed the bla_{TEM}-gene. PCR amplification and subsequent sequencing of a portion of these *bla*-genes revealed differences in produced beta-lactamases, especially for the CTX-family. CTX-M-1 was the most prevalent type; other alleles found included CTX-M-14 (LFMPF 181), CTX-M-56 (LFMPF 162, 270, and 274), CTX-M-35 (LFMPF 275) and CTX-M-15 (LFMPF 278).

5.5 Discussion

The fact that Belgian chicken meat preparations were found positive for *Campylobacter* spp. in 41.0 % of the examined samples after enumerations and only in 24.2 % of the same examined samples after enrichment (Habib *et al.*, 2008a) raised the question about suppression of *Campylobacter* spp. by other bacteria in the Bolton enrichment broth. Overgrowth of pathogenic target microorganisms during enrichment is often described but mostly for taxonomically related spp. e.g. *Listeria innocua* and *L. monocytogenes* competition (Besse *et al.*, 2005; Cornu *et al.*, 2002; Flanders *et al.*, 1995; Grant, 2004). It can be established from the present study that Bolton broth supported the growth of ESBL *E. coli* isolated from Belgian poultry meat and that these ESBL *E. coli* crowded the mCCDA plates resulting in overgrowth of *C. jejuni*. The results obtained in this study show true risk of obtaining false negative results when using ISO 10272:2006 (International Standards Organization, 2006b) for detection of *Campylobacter* spp. in poultry meat holding ESBL *E. coli* as was established before in our survey for *Campylobacter* spp. in chicken meat preparations in Belgium (Habib *et al.*, 2008a).

E. coli is reported not to grow in Bolton broth in the technical sheet of Bolton broth (Bridson, 2006), although Baylis *et al.* (2000b) reported high incidence of *E. coli* and *Pseudomonas* spp. in Bolton broth. In a preliminary test of co-cultures with a culture collection *E. coli* LFMFP 168 strain, indeed no interference of this *E. coli* strain was noted for *C. jejuni* detection after enrichment in Bolton broth. However, *E. coli* isolates in the present study taken from Belgian chicken meat were shown to be resistant to concentrations up to at least four times the concentration of sodium cefoperazone used in Bolton broth. Sodium cefoperazone, which is also present in the mCCDA plates (the recommended isolation medium described in ISO 10272:2006 ((International Standards Organization, 2006b)) is a third generation cephalosporin antibiotic. The cephalosporins are a class of beta lactam antibiotics. Resistance of *Enterobacteriaceae* to antibiotics, including cephalosporins is increasing in Europe and worldwide (Canton *et al.*, 2008; Livermore *et al.*, 2008; Machado *et al.*, 2008; Suzuki *et al.*, 2009; Wang *et al.*, 2008). This study shows that the *E. coli* strains isolated from chicken carcasses were resistant to at least four antibiotics, and interfere with the detection of *Campylobacter* spp. because of their resistance to the concentration of sodium cefaperazone used in Bolton broth and in mCCDA. Those ESBL producing *E. coli* are shown to be multi-resistant to 11 tested antibiotics used in this study. Therefore it is not so easy to optimize a cocktail of antibiotics to be used as selective supplement in enrichment media.

It was shown in this study that the use of the same antibiotics in an enrichment broth and in the subsequent selective plating medium is of concern. In the case of non-frozen *C. jejuni* cells in the co-culture assay, Bolton broth could be used as enrichment broth, but only in combination with *Campylobacter* spp. specific real-time PCR. Selective plating on mCCDA was not appropriate because the mCCDA plates were crowded with ESBL *E. coli* after Bolton broth enrichment, resulting in a true risk of overlooking the *Campylobacter* spp. colonies. A cefoperazone concentration of 32 µg/ml present in mCCDA plates was insufficient to inhibit the growth of these high numbers of ESBL *E. coli* strains. It might be recommended to evaluate alternative isolation media such as Preston, Skirrow, Butzler or the newly introduced chromogenic media recently brought to the market by various suppliers of media e.g. Oxoid, BioMérieux or Bio-Rad for their ability to inhibit the growth of these ESBL *E. coli*.

For *Campylobacter* spp., which are non-competing microorganisms and very fastidious, meaning that they can be easily overgrown by accompanying flora, another option is the use of Preston broth, as was recommended by the former version of the ISO 10272:1995 (International Standards Organization, 1995), the more selective enrichment broth for detection of the pathogen in food with a high microbial load of competing flora. This is not necessarily to be generalized for all food types, as Preston broth might be too selective in some situation where *Campylobacter* spp. are sub-lethally injured. Bolton broth may still be opportune to use as this enrichment broth will promote better resuscitation and growth compared to the more selective Preston broth as it is not clear accurately how widely distributed these ESBL *E. coli* are in other countries. To avoid the risk of false negative results perhaps, for a horizontal method applicable to all foods, a parallel enrichment in Bolton and Preston broth might be considered. The findings of ESBL *E. coli* present on poultry meat should be confirmed by other countries.

5.6 Conclusion

It can be established from the present study that Bolton broth supported the growth of ESBL *E. coli* isolated from Belgian poultry meat and that these ESBL *E. coli* crowded the mCCDA plates resulting in overgrowth of *C. jejuni*. The results obtained in this study show true risk of obtaining false negative results when using ISO 10272:2006 (International Standards Organization, 2006b) for detection of *Campylobacter* spp. in poultry meat holding ESBL *E. coli*.

E. coli isolates in the present study taken from Belgian chicken meat were shown to be resistant to concentrations up to at least four times the concentration of sodium cefoperazone used in Bolton broth. Sodium cefoperazone, which is also present in the mCCDA plates is a third generation cephalosporin antibiotic.

For *Campylobacter* spp., which are non-competing microorganisms and very fastidious, meaning that they can be easily overgrown by accompanying flora, another option is the use of Preston broth, as was recommended by the former version if the ISO 10272:1995 (International Standards Organization, 1995), the more selective enrichment broth for detection of the pathogen in food with a high microbial load of competing flora.

To avoid the risk of false negative results perhaps, for a horizontal method applicable to all foods, a parallel enrichment in Bolton and Preston broth might be considered. The findings of ESBL *E. coli* present on poultry meat should be confirmed by other countries.

Chapter 6

Appropriate sample preparation for reliable real-time PCR detection of low numbers of healthy and sub-lethally injured *Salmonella enterica* in chocolate

Redrafted after:

Vicky Jasson, Leen Baert and Mieke Uyttendaele

(2009)

Letters in Applied Microbiolgy, submitted

Chapter 6 Appropriate sample preparation for reliable real-time PCR detection of low numbers of healthy and sub-lethally injured *Salmonella enterica* in chocolate

6.1 Abstract

The aim of this study was to compare different enrichment and DNA extraction protocols to assess their capacity to detect low levels of healthy, but also sub-lethally injured *Salmonella enterica* cells in chocolate as food matrix with real-time PCR. Real-time PCR, iQ-Check^{TM} *Salmonella* II (Bio-Rad, Marnes-la-Coquette, France), was used as detection assay. VIDAS® Easy SLM (BioMérieux, Marcy l'Etoile, France) and ISO 6579:2005, the reference method, were used as confirmation methods. Both alternative methods are validated by AFNOR for all foods. In this study the food matrix chocolate, a low moisture food known to support the survival of *Salmonella* spp., was challenged. Two DNA extraction protocols, Standard and Easy extraction protocol (Bio-Rad), both applicable before iQ-Check^{TM} *Salmonella* II, were tested. Easy extraction protocol showed to be more appropriate as DNA extraction method for chocolate as food matrix. Buffered Peptone Water, the recommended ISO pre-enrichment medium was shown to be incapable to recover low levels of sub-lethally injured *S. enterica* cells from chocolate. The addition of milk powder to the BPW enhanced the ability to detect low numbers of *S. enterica*. All positive samples could be confirmed by another technique (VIDAS® or ISO 6579:2005 method). This was expected as it was shown that at least 250 CFU/ml of *S. enterica* should be present after enrichment to provide a positive PCR result with the protocols applied, similar to the other two methods. Rapid and reliable detection of *S. enterica* cells in chocolate by real-time PCR was proven to remain a challenge due to low numbers of sub-lethally injured target cells and potential inhibition of the PCR reaction.

6.2 Introduction

Salmonella spp. is a major causative agent of food-borne outbreaks in the European Union (EFSA, 2009). Many foods have been identified as vehicles for the transmission of *Salmonella* spp. to humans in particular meat (Urfer *et al.*, 2000), poultry, eggs (Gantois *et al.*, 2009), milk (Brouard *et al.*, 2007) and in lesser extent fruit and vegetables (Little and Gillespie, 2008). Also chocolate has been involved in food-borne outbreaks of *Salmonella* spp. (Cordier, 1994; Torresvitela *et al.*, 1995; Werber *et al.*, 2005). Food-borne outbreaks due to consumption of *Salmonella* spp. contaminated chocolate products appear to be associated with the ingestion of very small numbers (1 CFU/100g) of viable or recoverable cells of salmonellae (Komitopoulou and Penaloza, 2009). Reliable detection of low numbers of *Salmonella* spp. is therefore of the utmost importance. Detection of *Salmonella* spp. in foods occurs classical with the ISO 6579:2005 method (International Standards Organization, 2005b) which is based on a two-steps enrichment protocol followed by plating on selective agar plates and a final confirmation step. The ISO method is time-consuming posing a major drawback as positive results are obtained after 5-7 days. Therefore a wide range of alternative "rapid" detection method such as lateral flow devices, ELISA (Bohaychuk *et al.*, 2005; Croci *et al.*, 2001), ELFA (e.g. VIDAS®) (Vieira-Pinto *et al.*, 2007; Yeh *et al.*, 2002), PCR and real-time PCR methods (Krascsenicsova *et al.*, 2008; McMahon *et al.*, 2004; Uyttendaele *et al.*, 2003) have been developed over the years for *Salmonella* spp. detection (Eijkelkamp *et al.*, 2009).

Unfortunately these "rapid" methods, such as PCR, are not as "rapid" as they seem, because they also need the time-consuming resuscitation and enrichment steps. The PCR detection itself has become a reliable and acceptable detection technique for food-borne pathogens as the technique itself has grown out of one's infancy. Theoretically PCR can detect 1 cell per PCR reaction, yet due to the limited DNA input, a few hundred to thousands pathogenic cells are needed for the PCR reaction in order to reliable detect the presence of these pathogens (Hanna *et al.*, 2005; Rodriguez-Lazaro *et al.*, 2007). Overlooking or minimizing appropriate resuscitation and enrichment steps may result in false negative results especially when low numbers of sub-lethally injured cells are present (Osborne and Bremer, 2002). This is in particular the case for *Salmonella* spp. present in chocolate, a low moisture food, where the pathogen is subjected to a low water activity and although having the ability to survive for a prolonged period it may be sub-lethally injured. The advantage however is that *Salmonella* spp. are intrinsic fast growing microorganisms which overall do not need special nutritional requirements for enhanced growth. Buffered Peptone Water (BPW), a non-selective simple medium, is widely used and found appropriate for *Salmonella* spp. as prescribed in ISO 6579:2005 (International Standards Organization, 2005b).

Additionally sample preparation is important to avoid inhibition of the PCR reaction, which is frequently associated with fatty or protein rich matrices (Maurer J., 2006; Uyttendaele *et al.*, 2003). Nowadays commercially available real-time PCR kits usually incorporate an internal control in each reaction to exclude false negative results due to inhibition (Hoorfar *et al.*, 2003; Hoorfar *et al.*, 2004; Rijpens *et al.*, 2002).

The aim of this study was to compare different enrichment and DNA extraction methods to assess their capacity to detect low levels of healthy and sub-lethally injured *Salmonella enterica* cells in chocolate as food matrix by real-time PCR. The AFNOR validated iQ-Check™ *Salmonella* II PCR kit (Bio-Rad, Marnes-la-Coquette, France) served as real-time PCR format. The Standard and Easy extraction protocols, both recommended as potential DNA extraction methods by Bio-Rad for use with iQ-Check™ *Salmonella* II, were compared to provide sufficient and purified DNA for the PCR reaction. The AFNOR validated VIDAS® Easy SLM (BioMérieux, Marcy l'Etoile, France) and the ISO 6579:2005 reference method (International Standards Organization, 2005b) were used to confirm the results obtained by real-time PCR.

6.3 Materials and methods

6.3.1 Bacterial strains and culture conditions

A reference stock of the *Salmonella enterica* subsp. *enterica* strain LMG 10395 (*S.* Enteritidis) was kept at −70°C in Tryptone Soy Broth (TSB; Oxoid, Basingstoke, UK), supplemented with 0.6% yeast extract (YE; Oxoid), with 15% glycerol (Prolabo, Heverlee, Belgium). A stock culture was kept at 4°C on a Tryptone Soy Agar (TSA; Oxoid) slope supplemented with 0.6% YE. The stock culture was transferred monthly. Working cultures of *S. enterica* were prepared by loop inoculation in 10 ml of fresh TSBYE and incubated for 24 h at 37°C.

A standardized procedure providing a significant percentage of sub-lethal injury of *S. enterica* cells (≥ 98%) was obtained by freezing *S. enterica* in TSBYE for 7 days at –21°C. Percentage of sub-lethally injured cells was determined by plating on non-selective TSAYE and selective xylose lysine deoxycholate agar (XLD; Oxoid) as described earlier in Chapter 2.

6.3.2 DNA extraction protocols

Two DNA extraction protocols, Standard extraction protocol and Easy extraction protocol (Bio-Rad), described by the manufacturer as applicable for iQ-Check™ *Salmonella* II, were evaluated with chocolate as food matrix. For Standard extraction protocol, 1 ml of enrichment broth was taken and centrifuged for 5 min at 10000 x g, supernatant was discarded and 200 µl of the lysis reagent was added. The resuspended pellet (200 µl) was heated for 10 min at 100°C followed by a centrifugation step (5 min at 10000 x g). Five µl of the supernatant was taken for the subsequent PCR reaction. For the Easy extraction protocol, 100 µl of the enrichment broth was taken and 100 µl of lysis reagent was added before heating at 100°C for 10 min, followed by a centrifugation step (5 min at 10000 x g). Five µl of the supernatant was taken for the subsequent PCR reaction. In case real-time PCR reaction was inhibited, the extracted DNA was diluted 1/10 in RNase/DNase free water (Qiagen, Hilden, Germany) and samples were reanalyzed.

6.3.3 Real-time PCR using iQ-Check™ *Salmonella* II (Bio-Rad)

iQ-Check™ *Salmonella* II is a molecular beacon based real-time PCR test and has obtained AFNOR validation certificate (certificate number: BRD 07/06–07/04) for all foods. Five µl of DNA extract (see above) was used for the amplification reaction (45 µl PCR mix). Initial denaturation, 10 min 95°C, was followed by the amplification consisting of 50 cycles with 15 seconds of denaturation at 95°C, 20 seconds of annealing at 58°C (detection step) and 30 seconds of extension of DNA at 72°C. Internal controls were included in each PCR reaction as well as 1 negative and 1 positive control per run (supplemented by the manufacturer). Samples were run on the iCycler thermal cycler (Bio-Rad).

The intrinsic detection limit of iQ-Check™ *Salmonella* II PCR technique based on Standard and Easy extraction protocol was examined. Therefore, chocolate, obtained from a local producer (25 gram), was aseptically taken and transferred into a sterile filter stomacher. Chocolate was diluted 1/10 in Buffered Peptone Water and spiked with different levels of *S. enterica* cells (250, 25 en 0 CFU/ml). Spiking was performed with healthy cells. Subsequently, aliquots of 1 ml or 0.1 ml were immediately taken for DNA extraction (no prior enrichment applied) using respectively Standard and Easy extraction protocol. Similarly *S. enterica* inoculated samples of BPW without chocolate served as control samples to check for any potential matrix effect. Experiments were performed in duplicate.

Next the detection limit of the PCR method (including enrichment step) was evaluated. The effect of enrichment phase of *S. enterica* on the DNA extraction protocols to provide purified DNA for an efficient PCR reaction was examined. In particular the effect of incubation of chocolate in BPW during the enrichment phase was tested. Therefore chocolate was spiked with low numbers (approximately 10 CFU/25g) of healthy *S. enterica*. Next samples (0.1 ml and 1 ml) were taken for the corresponding DNA extraction protocols after enrichment (21 h at 37°C) and PCR was performed.

6.3.4 Evaluation of real-time PCR detection for reliable detection of low numbers of *S. enterica* in chocolate using VIDAS® and ISO 6579:2005 as confirmation techniques

A 50 gram chocolate sample was aseptically taken and transferred into a sterile filter stomacher bag. A ten fold dilution was made in enrichment broth and the sample was homogenized during 60 s by means of a stomacher. "Low" levels (1-5 CFU/50g) as well as "high" levels (5-50 CFU/50g) of healthy and sub-lethally injured *S. enterica* cells were used to inoculate the chocolate samples. Control samples were inoculated with sterile PPS (blanks). Enrichment was performed in either Buffered Peptone Water (BPW; Oxoid) (21 ± 1 h at 37°C) or in Buffered Peptone Water supplemented with 100 g/l sterile dried skimmed milk powder (Oxoid) and 0.018% Brilliant Green (BPW+SM+BG) as commonly used in the chocolate industry. Enrichment in BPW+SM+BG consisted of an enrichment in BPW+SM at 37°C for 2 h followed by the addition of brilliant green and further incubation for 19 ± 1 h at 37°C. All experiments were performed in triplicate. After enrichment samples were taken for Easy extraction protocol and subsequent PCR or appropriate aliquots were taken to run the conventional ISO 6579:2005 or VIDAS® Easy SLM method as described below.

6.3.4.1 VIDAS® Easy SLM (BioMérieux)

Detection by VIDAS®, an automated Enzyme linked fluorescent assay (ELFA) which has different protocols such as VIDAS® Easy SLM, and which is validated by AFNOR (certificate number: BIO 12/16–09/05) for all foods, was performed. Briefly, 100 µl of enriched sample (BPW or BPW+SM+BG for 20 hours) was taken and inoculated in 10 ml of *Salmonella* Xpress broth (SX; BioMérieux) for 22-26 h at 41.5°C. Next *Salmonella* Xpress broth was heated for 15 min at 131°C and 0.5 ml was taken for detection with VIDAS®.

6.3.4.2 ISO 6579:2005

After enrichment in BPW or BPW+SM+BG for 20 h 100 µl was taken and transferred into 10 ml of Rappaport-Vassiliadis Soya Peptone broth (RVS; Oxoid) for further enrichment (24 h at 41.5°C). Besides, 1 ml was taken and transferred to 10 ml of Muller-Kauffmann tetrathionate-novobiocine broth (MKTTn; Oxoid) for 24 h at 37°C. Presence of *S. enterica* cells was isolated after incubation by plating on XLD and ChromID *Salmonella* (BioMérieux) medium for both broths.

6.4 Results

6.4.1 Procedure to for sub-lethally injured *S. enterica* cells

Inoculated TSBYE was stored for 7 days in a freezer at -21°C. After 7 days of freeze treatment, the cultures were thawed in the refrigerator at 4°C and plated in parallel on selective (XLD) and non-selective media (TSBYE). Table 6.1 shows the results obtained for 6 repetitions.

Table 6.1. Counts of *S. enterica* cells on TSAYE and XLD (before and after freezing) and the percentage of sub-lethal injury.

Repetition	Log CFU/ml before treatment		Log CFU/ml after treatment		% sub-lethal injury
	TSAYE	XLD	TSAYE	XLD	
1	5.99	6.08	5.26	3.72	97.06
2	6.04	5.90	5.20	3.49	98.06
3	6.01	5.91	5.48	3.43	99.10
4	5.92	5.89	5.38	3.20	99.33
5	5.89	5.67	5.05	2.45	99.75
6	5.74	5.75	5.04	3.34	98.00

Results of a two sided T-test (α = 0.05) showed no significant difference between counts on XLD and on TSAYE before treatment (p value = 0.2040). The threshold value (calculated as described in section 2.4.1 of Chapter 2) obtained for XLD was 30.1%. The mean % of sub-lethal injury imposed due to freezing showed to be 98.55 ± 1.01 %. The mean percentage of sub-lethal injury was shown to be significant different compared to the threshold value (one sided T-test, α = 0.05, $p < 0.0000$).

6.4.2 Identification of most appropriate DNA extraction protocol and the detection limit of iQ-Check™ *Salmonella* II in chocolate

Two DNA extraction methods were compared for their capacity to provide sufficient and purified DNA for PCR reaction. Table 6.2 shows that independently from the amount of cells spiked, DNA extraction with Standard extraction protocol resulted in inhibition of the PCR reaction in 7 out of 8 samples. After diluting the DNA and reanalyzing, the samples were found positive by real-time PCR if 250 CFU/ml or more cells were present.

Table 6.2. Detection of low numbers of *S. enterica* cells in chocolate with iQ-Check™ *Salmonella* II (Bio-Rad) based on Standard and Easy extraction protocol with and without enrichment.

Extraction protocol	Spiking (CFU/ml)	Enrichment	Copies for PCR reaction/5 µl[(a)]	Chocolate rep 1[(b)]	Ct value Chocolate rep 2[(b)]	No chocolate
Standard	250	No	6.25	Inh[(c)]/40	Inh/39	37
	25	No	0.625	Inh/-[(d)]	Inh/-	40
	0	No	0	Inh/-	Inh/-	-
	13 CFU/250 ml	Yes	NA[(e)]	25	Inh/28	17
Easy	250	No	0.625	43	44	39
	25	No	<detection limit	-	-	-
	0	No	0	-	-	-
	13 CFU/250 ml	Yes	NA	24	38	19

(a): e.g. for Standard extraction: 250 CFU/ml -> take 1 ml and concentrate until 0.2 ml -> lysis -> take 5 µl out of 0.2 ml for PCR reaction -> 6.25 CFU/ PCR reaction in theory

e.g. for Easy extraction: 250 CFU/ml -> take 0.1 ml sample and add 0.1 ml lysis reagent -> lysis -> take 5 µl out of 0.2 ml for PCR reaction -> 0.625 CFU/ PCR reaction in theory

(b): rep = repetition

(c): Inh = Inhibition. If inhibition of PCR reaction occurred, the sample was 1/10 diluted and reanalyzed.

(d): - = Ct > 50

(e): NA = not applicable

For the control samples (no chocolate) Ct values of 37 and 40 were obtained respectively when 250 and 25 CFU/ml *S. enterica* was present. DNA extraction with Easy extraction protocol resulted in detection of *S. enterica* in the case 250 CFU/ml was present (without the need for 1/10 dilution). A shift in Ct value from 39 in the control sample towards 43 and 44 in the samples with chocolate (correlated with an inoculation level of 250 CFU/ml) could be noticed, indicating an effect of chocolate on the PCR reaction. The results presented in Table 6.2 show that in the case of chocolate as food matrix at least 250 CFU/ml should be present to obtain a PCR signal.

6.4.3 Evaluation of real-time PCR detection for reliable detection of low numbers of *S. enterica* in chocolate using VIDAS® and ISO 6579:2005 as confirmation techniques

Table 6.3. Detection of *S. enterica* in artificially contaminated chocolate samples. Detection occurred with iQ-Check™ *Salmonella* II, VIDAS® Easy SLM and ISO 6579:2005.

Inoculum (cells/50g)	Physiological state inoculum	Broth	Repeats	Detection method: RT-PCR (Ct value ± stdev)	VIDAS (Test value)[(a)]	ISO method: RVS	ISO method: MKTTn
20	Healthy	BPW	1	20.9 ± 0.1	2.98	+	+
		BPW	2	25.1 ± 0.3	2.95	+	+
		BPW	3	26.1 ± 0.5	2.93	+	+
		BPW+M+BG	1	27.9 ± 0.7	3.15	+	+
		BPW+M+BG	2	32.7 ± 0.4	3.14	+	+
		BPW+M+BG	3	29.7 ± 2.0	3.18	+	+
2	Healthy	BPW	1	23.1 ± 0.1	3.13	+	+
		BPW	2	27.0 ± 0.2	3.02	+	+
		BPW	3	28.6 ± 0.4	2.96	+	+
		BPW+M+BG	1	33.7 ± 1.0	2.93	+	+
		BPW+M+BG	2	38.0 ± 1.0	3.17	+	+
		BPW+M+BG	3	33.7 ± 0.6	3.19	+	+
40	Sub-lethally injured	BPW	1	27.1 ± 0.2	2.85	+	+
		BPW	2	26.6 ± 0.6	2.89	+	+
		BPW	3	23.1 ± 0.1	2.91	+	+
		BPW+M+BG	1	24.4 ± 0.1	2.76	+	+
		BPW+M+BG	2	31.0 ± 2.6	2.74	+	+
		BPW+M+BG	3	37.0 ± 3.5	2.79	+	+
4	Sub-lethally injured	BPW	1	-	-	-	-
		BPW	2	-	-	-	-
		BPW	3	-	-	-	-
		BPW+M+BG	1	32.3 ± 0.1	2.76	+	+
		BPW+M+BG	2	-	2.73[(b)]	-[(b)]	+[(b)]
		BPW+M+BG	3	25.6 ± 0.5	2.76	+	+
0	Blanks	BPW	1	-	-	-	-
		BPW	2	-	-	-	-
		BPW	3	-	-	-	-
		BPW+M+BG	1	-	-	-	-
		BPW+M+BG	2	-	-	-	-
		BPW+M+BG	3	-	-	-	-

(a): VIDAS@ is found positive if test value is >= 0.23

(b): Discrepancy between results due to second enrichment. Second enrichments were also analyzed with iQ-Check™ *Salmonella* II and were found positive.

-: No detection of *S. enterica* cells

+: *S. enterica* present on agar plates

Table 6.3 shows the results of detection of *S. enterica* cells in chocolate samples obtained with 3 different detection techniques. Healthy cells but also sub-lethally injured cells of *S. enterica* were analyzed after enrichment in BPW and BPW+SM+BG. For iQ-Check™ *Salmonella* II detection sampling and DNA extraction was performed in duplicate to evaluate variation due to sampling and DNA extraction.

Firstly, from Table 6.3 it can be noticed that all samples spiked with healthy cells were found positive independently from detection method or enrichment broth. Ct values of parallel sample were always higher when BPW+SM+BG was used as enrichment medium compared to BPW. Secondly, in case low levels of sub-lethally injured *S. enterica* were spiked no growth of the bacteria occurred when BPW was used as enrichment medium. Although when BPW+SM+BG was used as enrichment medium, 2 out of 3 samples tested positive independent from the detection method. However, in one of the repetitions, no PCR signal could be obtained with BPW+SM+BG as enrichment broth while VIDAS® and ISO 6579 using MKTTn, allowed detection after second enrichment. To confirm these results DNA was extracted from second enrichment in SX (VIDAS®), RVS or MKTTn (ISO 6579:2005) with Easy extraction protocols and resulted in PCR detection. Ct values of RVS (Ct = 36) were significant higher compared to Ct values of MKTTn (Ct = 20) and SX (Ct = 26).

6.5 Discussion

Recovery of food-borne pathogenic bacteria from food and feed or from the processing environment has traditionally been a challenge. Relative low levels of pathogens (1-100 CFU) can be present in large amount of food (often 1-100 g used for food analysis) while only a small volume size (10-500 µl) is used as input for the detection technique. Also the presence of sub-lethally injured bacteria (Kell *et al.*, 1998; Wu, 2008) and the non-uniform distribution of pathogens in food hamper the detection. Although some detection techniques have the intrinsic capacity to detect as little as one cell, the fact that the detection takes place from food increases the detection limit significantly. Food components such as some kinds of proteins and fat are often described to inhibit the PCR reaction (Malorny *et al.*, 2003b; Rodriguez-Lazaro and Hernandez, 2006). Also cacao is described as an inhibitory compound for PCR (Gryson *et al.*, 2007; Porebski *et al.*, 1997), due to its high amount of polyphenols and polysaccharides (Taq polymerase inhibitors), making chocolate a difficult food matrix for PCR. In this study it was shown that if Standard extraction protocol (Bio-Rad) was applied for chocolate as food matrix, in almost all cases inhibition of the PCR reaction occurred. However it was possible to detect *S. enterica* after diluting the DNA, extracted with Standard extraction protocol. It was shown in this study that Easy extraction protocol (Bio-Rad) was more appropriate for DNA extraction in chocolate samples as fewer samples were inhibited. The explanation for lack of inhibition with Easy extraction protocol can be found in the volume of samples taken (0.1 ml versus 1 ml) and thus corresponding lower levels of inhibitory compounds taken that might be co-extracted.

A shift towards higher Ct values due to the presence of chocolate was shown in this study. The chocolate matrix increased the Ct values to similar values as obtained with samples almost reaching the critical theoretical limit of detection for PCR. PCR detection in the presence of chocolate resulted in 4 to 5 Ct values higher (if 250 CFU/ml of *S. enterica* were present, which resulted in theoretically ± 1 copy for the PCR reaction). Similarly Gryson *et al.* (2004) observed higher Ct values due to the presence of cacao and increased therefore the amount of cycles. This study showed that for detection of *S. enterica* in chocolate at least 250 CFU/ml of *S. enterica* should be obtained after enrichment to provide enough DNA copies in the sample volume to come up with a positive PCR result. Similar results were obtained by Tortajada *et al.* (2009) who compared different DNA extraction protocols for microorganisms from cacao extracts and showed a detection limit of at least 100 CFU/g.

Due to the low infectious dose of *Salmonella* spp. and the fact that real-time PCR in combination with Easy extraction needed a minimum of 250 CFU/ml to avoid false negative results, fast and reliable enrichment is needed. It has been shown, in Chapters 3-5, and by others (Osborne and Bremer, 2002; Uyttendaele *et al.*, 2001a) that enrichment broths differ significantly in the ability to resuscitate injured cells. Influence of enrichment broth on the recovery of low numbers sub-lethally injured *S. enterica* cells was shown in this study (Table 6.3). Low numbers of sub-lethally injured *S. enterica* cells could not be recovered if BPW was used as enrichment broth, while BPW+SM+BG was capable to recover sub-lethally injured *S. enterica* in 2 out of 3 samples independent from the detection method. This showed that skim milk powder can be used as a growth enhancing factor.

If *S. enterica* was detected with real-time PCR method, at all times, positive results could be confirmed by classical plating or ELFA. This can be expected if for a positive PCR result 250 CFU/ml are needed, these numbers of cells are also known to be sufficient to be picked up by a secondary enrichment broth (using 0.1 to 1 ml from pre-enrichment to transfer). Although PCR can be regarded as a more rapid test, it is, as a method, not more sensitive, but rather equivalent compared to the other alternative detection methods. As such it is shown that the main bottleneck for PCR, a method susceptible to food interference testing small volumes, is an appropriate sample preparation. In one repetition out of three, low levels of sub-lethally injured cells needed a second enrichment step to growth until detectable numbers, because in case of real-time PCR immediately after BPW+SM+BG no detection occurred while VIDAS® and ISO 6579:2005 MKTTn allowed detection after second enrichment. These results are in accordance with (Fakhr *et al.*, 2006) who found similar results for *Salmonella* in naturally contaminated turkey.

It should be noted that the experiments performed in this study should represent a worst case scenario as very low numbers of maximum sub-lethally injured *S. enterica* cells were used (98 ± 1 % sub-lethal injury) which are only expected in particular food types which present adverse conditions for *Salmonella* to persist.

6.6 Conclusions

It was shown in this study that detection of *S. enterica* from chocolate by real-time PCR is a challenge due to on the one hand the presence of low levels of sub-lethally injured cells and on the other hand due to the inhibition of the PCR reaction. BPW failed to recover low levels of sub-lethally injured *S. enterica* cells from chocolate while BPW supplemented with skim milk powder allowed resuscitation and enrichment of these low numbers of sub-lethally injured cells.

It might be recommended to prolong the enrichment step or to include a short selective enrichment/concentration step before PCR is applied to improve the sensitivity of the test to avoid false negative results.

Chapter 7

Bottlenecks and selection criteria for rapid methods in foods: General discussion, conclusion and perspectives

Partially redrafted after

Vicky Jasson, Liesbeth Jacxsens, Pieternel Luning, Andreja Rajkovic and Mieke Uyttendaele

(2009)

Food Microbiology, submitted

Chapter 7 Bottlenecks and selection criteria for rapid methods in foods: General discussion, conclusion and perspectives

7.1 Resuscitation a prerequisite for rapid detection of food-borne pathogens

Food-borne illnesses are of concern due to their impact on both humans and economy. *L. monocytogenes* is a relevant food-borne pathogen due to the high mortality rate (20%) of infected people, while *Campylobacter* spp. and *Salmonella* spp. are relevant due to their high frequency of causing illness (respectively 55.8 and 37.5 annual confirmed cases/10000 persons in Belgium)(EFSA, 2009). *E. coli* O157:H7 is relevant due to both a low infectious dose and the severity of the associated illness (Chapter 1). Microbiological analyses regarding these “big 4” are important in different niches such as governmental food inspection, in commercial relationships, in the food industry, in academic laboratories and in reference laboratories. The numbers of reported outbreaks (EFSA, 2009), the impact on society, the low infectious dose of these food-borne pathogens and the severity of the illnesses, imply the need for sufficient sensitive detection methods for these microorganisms in foods (1-10 CFU/25g).

Standardized methods (e.g. ISO methods) are acknowledged as the reference analytical methods for official control. These standardized methods are most of the time classical culture methods, using selective nutritious broth or agar media to grow, isolate or enumerate the target organism while suppressing the indigenous background flora of the food. They have been developed historically based upon expertise and selected by international consensus. These standardized classical culture methods are harmonized methods, looked at as the “gold standards” in food diagnostics and thus overall well accepted (Feng, 1996; Fung, 1994). A serious draw back is that, although they demand no expensive infrastructure and are rather cheap in consumables, they are laborious to perform, demand large volumes of liquid and solid media and reagents, and encompass time consuming procedures both in operation and data collection.

Fresh and minimal treated foods are gaining popularity by the consumer due to their healthy image. These fresh foods have a short shelf-life. With conventional detection methods the time required to ensure the absence of pathogens would be excessive. In addition both food business operators and competent authorities are urged to act quickly and implement corrective actions if food safety is threatened. This enforces the desire of the use of rapid alternative detection methods. In **Chapter 1** an overview is given regarding the principles and elaboration of various alternative "rapid" detection methods.

Although in most of the cases, the aim of these alternative detection methods, as described in Chapter 1, is to reduce the time to obtain a result, still the first **prerequisite** is the **reliability of the method**. Because food processing (freezing, heat-treatment, etc.) and food preservation techniques (salting, acidification, packaging, etc.) present non-optimal conditions for microorganisms, the pathogens in the food are stressed and/or sub-lethally injured. Appropriate resuscitation and enrichment steps can avoid false negative results due to not reaching a certain detection limit for a certain detection technique in a certain time. It is therefore of the utmost importance to incorporate sub-lethally injured cells for testing performance of resuscitation and enrichment media. Therefore one of the objectives of this PhD dissertation was to determine standardized procedures for creating **sub-lethally injured cells**, as described in **Chapter 2**.

The experiments presented in Chapter 2 demonstrated that it **was feasible** to set up **protocols provoking injury** to cells in a controlled reproducible manner enabling simulations of the status of the natural flora of naturally contaminated samples. Inclusion of these obtained injured cells when assessing performance of the resuscitation and enrichment broth for their recovery capacities better reflect the real life situation and take into account the history of the cells and the degree of sub-lethal injury of the natural contamination. To determine standard methods for creating sub-lethally injured cells, the different pathogens were subjected to different types of stress related to food processing and storage. Threshold values were determined to obtain significant percentages of sub-lethally injured cells created by a defined stress condition which was different from other studies (Besse *et al.*, 2000; Dykes and Withers, 1999; Hara-Kudo *et al.*, 2000; Raccach *et al.*, 2002). To our knowledge no standardized procedures were described yet including a threshold value to determine the significance of the percentage of sub-lethal injury for different pathogens. Also strain variation for the different pathogens was taken into account while establishing these protocols. It was shown in Chapter 2 that depending on the microorganism **different stress situations** were capable of creating a significant amount of sub-lethal injury. It was also noticed in Chapter 2 that if presence/absence testing is performed in the case very low numbers are present, the sub-lethally injured fraction may become a very important part of the population of low numbers. Also if low numbers are present (<100 CFU/ml) and high percentages of sub-lethal injury are noted (>90% as noted for *E. coli* O157) the number of uninjured cells is reduced to the single cell level and variation in lag-phase distribution becomes important as well (Francois *et al.*, 2006; Stephens *et al.*, 1997). In these cases, resuscitation of the sub-lethally injured cells will need a prolonged incubation time or a prior incubation in a non-selective medium to ensure detection. This phenomenon can partly explain the failure of some (advanced) detection methods to detect low numbers of pathogens in processed foods if the enrichment protocol is too short or does not allow appropriate resuscitation of injured cells. This points out the need for including sub-lethally injured cells into the validation studies in order to assess the potential of conventional cultural methods, as well as, advanced detection methods to reliable detect and/or enumerate food-borne pathogens in foods. As naturally contaminated foods are not always available, the procedures for provoking injury established in this study for each of the pathogens involved may aid in establishing the detection limit and robustness of the rapid screening methods for pathogens in food diagnostics.

7.2 Four pathogens, four different stories of enrichment

It has been shown that enrichment broths differ significantly in the ability to resuscitate injured cells (Osborne and Bremer, 2002; Uyttendaele *et al.*, 2001a). In **Chapter 3** different selective and non selective broths were evaluated for their capacity regarding the recovery of healthy but also sub-lethally injured ***L. monocytogenes*** cells. Studies have been accomplished during the past years comparing different enrichment protocols or various enrichment broths (Balamurugan *et al.*, 2006; Beumer *et al.*, 1996; Grant, 2005; Lionberg *et al.*, 2003) but few of them looked in detail at the growth kinetics of the target bacterium in the broth (Duffy *et al.*, 1994a; Duffy *et al.*, 1994b; Stephens *et al.*, 1997). An extensive study, as presented in Chapter 3, of the kinetics during resuscitation and enrichment was not yet performed. In this Chapter 3 focus was set on the determination of $\boldsymbol{\mu_{max}}$ and $\boldsymbol{\lambda}$ for *L. monocytogenes* in various broths. From Chapter 3 it became clear that a **non selective broth** although very promising *in vitro* was **not useful** for the recovery of ***L. monocytogenes*** from foods. Despite the high μ_{max} and the short lag phase, enrichment using TSB failed to detect *L. monocytogenes* in approximately 50% of the analyses performed in artificially contaminated foods with both phage protein based capture/plating or VIT-*Listeria* Kit (FISH). The study showed that a **moderate selective enrichment medium**, such as LX or LSB is more useful to recover *L. monocytogenes* from foods. In combination with the VIT-*Listeria* Kit a 24 hours detection method is obtained (Stephan *et al.*, 2002; Stephan *et al.*, 2003). In comparison with ISO 11290 (International Standards Organization, 1996) where demi-Fraser and Fraser are used as enrichment broths, the use of LX or LSB would result in a reduction of 24 to 48 h in the enrichment step.

A similar experimental setup was used for ***E. coli* O157** in **Chapter 4** as indicated for *L. monocytogenes* in Chapter 3. μ_{max} and λ for healthy, stressed or sub-lethally injured *E. coli* O157 in various non selective and selective broths were determined. Limited enrichment broths were selected based on μ_{max} and λ and further evaluated with artificially contaminated raw beef, soy sprouts, raw milk and fermented sausage. The results obtained in Chapter 4 show that when using the ISO 16654 method (International Standards Organization, 2001a), false negative result can be obtained if low numbers of sub-lethally injured cells are present in food samples. **The selective mTSB+N**, used in the ISO 16654, showed to be an **inadequate** resuscitation broth to detect **sub-lethally injured** *E. coli* O157 cells. Using a **non selective enrichment broth** such as **BPW** revealed **positive results** and would therefore be **recommendable**.

As ***Campylobacter* spp.** is a fastidious microorganism prone to be overgrown by competitive indigenous flora, the use of a non-selective broth is not feasible. In a former version of ISO 10272 (International Standards Organization, 1995) the more selective Preston broth was recommended as enrichment medium for the detection of *Campylobacter* spp. and is still used in some countries (Baylis *et al.*, 2000b; Paulsen *et al.*, 2005). In 2006 the ISO 10272 was revised and Bolton broth was recommended as enrichment broth (International Standards Organization, 2006b). The change of broth occurred as Bolton broth had proven to be more appropriate in the recovery of injured cells (Baylis *et al.*, 2000b). Now, 3 years after the change it has become clear that some problems occur with the use of Bolton broth as enrichment medium. In Chapter 5 it was demonstrated that the use of **Bolton broth** can result in **false negative result** by the presence of **ESBL producing *E. coli*** in the food sample. Sodium cefoperazone, which is also present in the mCCDA plates (the recommended isolation medium described in ISO 10272:2006 ((International Standards Organization, 2006b)) is a third generation cephalosporin antibiotic. The cephalosporins are a class of beta lactam antibiotics. Resistance of *Enterobacteriaceae* to antibiotics, including cephalosporins is increasing in Europe and worldwide (Canton *et al.*, 2008; Livermore *et al.*, 2008; Machado *et al.*, 2008; Suzuki *et al.*, 2009; Wang *et al.*, 2008). The less selective enrichment Bolton broth followed by plating on mCCDA has shown to be not appropriate for the recovery of *Campylobacter* spp. from food sample contaminated with ESBL producing *E. coli.*

Especially for ***Campylobacter* spp.** it seems to be a **difficult compromise** because on the one hand it is preferred to use a non-selective broth to stimulate the **resuscitation of sub-lethally injured** cells while on the other hand selective compounds are required to **prevent overgrowth** by other flora. The fact that Bolton broth is less selective and allows faster growth of *C. jejuni* compared to Preston broth does not provide any advantage if competing flora is present, in particularly the antibiotic resistant *E. col*.

Taking into account the experience with previous 3 pathogens, experiments were set up to test the capacity of real-time PCR (iQ-check *Salmonella* II) to detect healthy and sub-lethally injured ***S. enterica*** cells from a defined food type namely **chocolate**. The food matrix was chosen because of its complexity and the high probability of the presence of low numbers and sub-lethally injured *Salmonella* spp. cells and results were presented in **Chapter 6**. In this experimental approach searching for the edge of resuscitation was possible. *Salmonella* spp. is recognised as fast and easy growing microorganism needing little nutrients. The ISO 6579 method (International Standards Organization, 2005b) recommends the non selective BPW as enrichment broth. To favour the resuscitation of *S. enterica* from chocolate skim milk powder was added to BPW (Chapter 6). It has been show that for raw beef an enrichment of 8 h in BPW was sufficient to recover the pathogen (Patel and Bhagwat, 2008; Warren *et al.*, 2007). It should be noted that in raw beef as food matrix the probability of the presence of sub-lethally injured cells is very small. In Chapter 6 it was show that this story does not hold for chocolate as this food presents harsh conditions for sub-lethally injured *Salmonella* spp. Sub-lethally injured *S. enterica*. in chocolate required at least 24 h of enrichment are needed.

It can be noticed from the results obtained in the Chapters 3-6 that **major differences** exist in the elaboration of a reliable detection strategy, in particular the **resuscitation step**, which is in the first place based on the **microorganisms themselves** (Figure 7.1). In case of *E. coli* O157, a very competitive microorganism and a fast grower only few nutrients are required and a basic non-selective broth, such as BPW, can be used for resuscitation and subsequent growth of injured *E. coli* O157 cells (Chapter 4). Similar, *Salmonella* spp. (Chapter 6) is also characterized by a robust nature and does not require special nutrient in the enrichment broth but addition of skim milk powder helps in the recovery of injured cells from chocolate. *L. monocytogenes*, an intermediate grower and more fastidious relatively, needs to be enriched in a selective broth with limited additions of selective component (Chapter 3). In the case of *C. jejuni* it can be claimed that potential overgrowth by other bacteria is as important as the resuscitation of the injured cells. *C. jejuni* is a non competing microorganism and very fastidious, easily overgrown by accompanying flora and thus requiring a selective medium (Chapter 4). In addition it should be noticed that the time needed for enrichment will vary with the type of microorganisms, but also with the **type of food** and the **stress** it exerts on the microorganisms present. In the particular case of e.g. *E. coli* O157 or *S. enterica* detection in e.g. fresh meat and fresh produce an enrichment of 5 to 8 h can be sufficient but in other types of foods, often food with shelf-life exceeding 10 days, at least an overnight and even up to 24 h is needed e.g. *Salmonella* in low moisture foods (Kotzekidou, 1998).

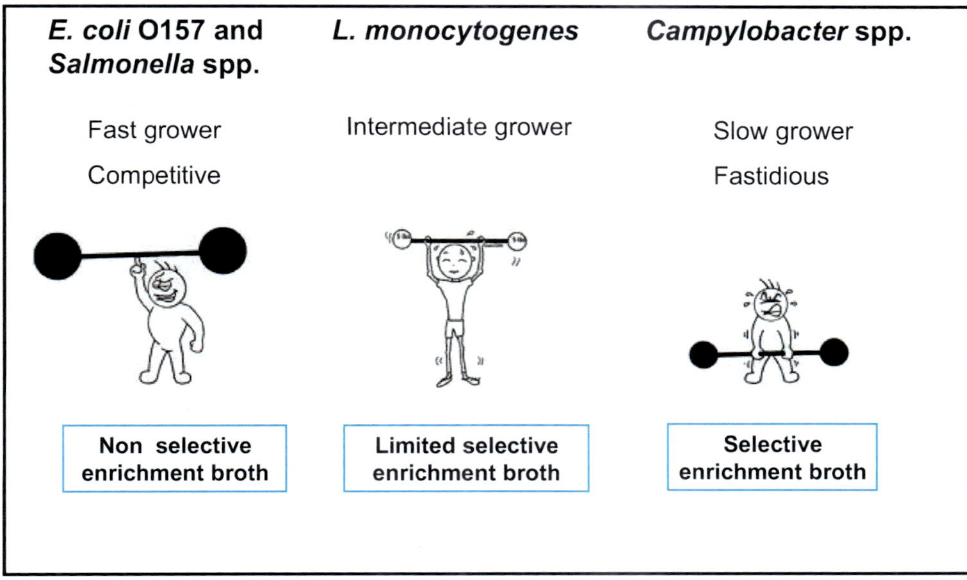

Figure 7.1. Comparison of *E. coli* O157, *Salmonella* spp., *L. monocytogenes* and *Campylobacter* spp. regarding their growth and resuscitation characteristics.

7.3 Selection criteria regarding alternative detection methods

For official controls the ISO method and reference methods are to be used. An important drawback of these ISO methods is the time required to obtain a microbiological result. In the case microbial analyses are performed for trend analyses to serve baseline studies or for monitoring purpose on CCP's, the time to obtain results is not necessarily of primary importance. In the case of an alert time is a crucial factor and rapid detection methods are needed. An overview of fast detection methods, in particular the validated methods (Table 1.3 in Chapter 1), is given in Chapter 1. Regulation 2073/2005 mentions indeed that alternative methods can be used if they have obtained a validation certificate based on ISO 16140 (International Standards Organization, 2003b). In deciding which methods to apply, there are more and also important criteria besides a validation certificate and the time to obtain results.

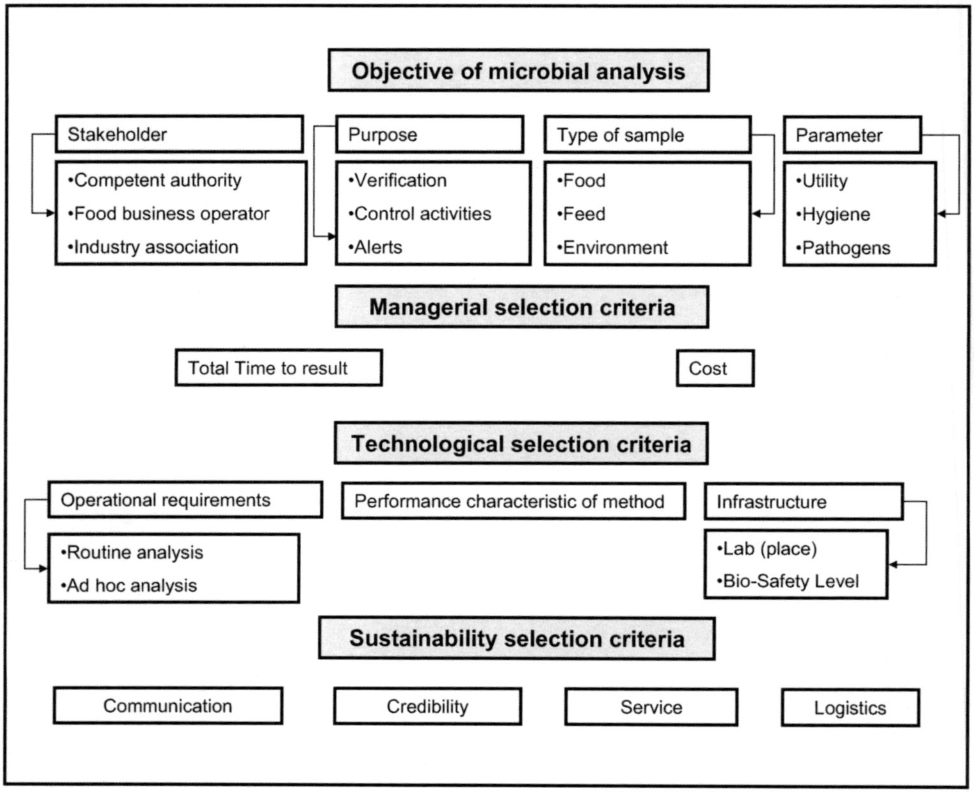

Figure 7.2. The decision tree for classifying microbial analysis in the framework of food safety and food quality.

Due to the overload of alternative methods and/or formats on the market, food business operators or competent authority, for which microbial analysis (MA) of food is only a supporting tool in the assurance of food safety, have difficulties in deciding which method is best fit for their purpose in their particular context. Figure 7.2 gives an overview of **different selection criteria** important in **consideration** for

alternative detection methods. These selection criteria can aid in the process of decision-making regarding the selection of the most appropriate methods for MA in a specific situation. The list of selection criteria can help the end user of the method to obtain a systematic insight into all relevant factors, beyond the technical performance characteristics of the method.

The selection criteria are chosen based on a **techno-managerial** point of view (Figure 7.2) and consists of four main inputs of information, I) objective of MA, II) managerial selection criteria, III) technological selection criteria and IV) sustainability selection criteria. The selection criteria can be linked with certain parameters belonging to the alternative methods and are discussed in the next section.

- **The objective of microbial analysis**

The first selection criterion deals with the objective for MA and contains basic information on the context of the MA. It requires input on the I) type of stakeholder, II) the purpose of microbial analysis, III) the type of sample and IV) the microbial parameter (type of microorganisms) of concern.

In the food supply chain three types of stakeholders can be distinguished who are frequently involved in MA: _competent authorities_, _industry associations_ or _food business operators_. Each of them might be bound internally or externally to recommendations or strict laws with regard to the selection of methods. The type of stakeholder may restrict the selection if only defined and/or **validated methods** are accepted by third parties judging the results (e.g. official control according to EU regulation 2073/2005 is demanding specific analyzing techniques).

Regarding the purpose of MA, also three types can be distinguished with increasing need for rapid action (and thus rapid methods): _monitoring_, _control activities_ or _alerts_. In case of a competent authority the alert can be an announcement of the presence of a hazard in a certain food e.g. *E. coli* O157 outbreak in The Netherlands due to raw beef (Greenland *et al.*, 2009), but in case of a food business operator an alert can be a complaint from consumers or a critical control point (CCP) that went out of control. Monitoring by a food business operator can be the verification of the performance of its Food Safety Management System (FSMS) by e.g. Microbial Assessment Scheme (MAS) (Jacxsens *et al.*, 2009), while for the competent authority monitoring can be baseline surveys to reveal trends to establish sources of food-borne pathogens in the food chain or in the frame of control activities. The purpose of the MA has reflection on the method parameter **total time to result**, but also on the method parameter **validation certificate**. In the case of an alert, e.g., a CCP that went out of control, time is a crucial factor, while in the case of baseline surveys, time is far less critical. Concerning the validation certificate, if verification of a HACCP plan takes place, an in-house method might be used, but in the case of MA in the framework of official analysis or import/export, methods should be comparable and internationally accepted, meaning being subjected to independent validation.

Also the type of sample itself is important (raw versus processed or animal versus plant origin). The type of food will affect the performance of the methods i.e. the potential for presence of interfering compounds. The commodity (fish, meat, dairy, etc.) and the applied processing (raw, fresh, dried, heat treated etc.) will also affect the types and levels of microorganisms. The type of sample reflects toward the method parameter **scope of the method**. Some food matrices can contain components that can for example inhibit the PCR reaction and put higher demands on sample preparation. Some food matrices can be highly contaminated with indigenous flora, with similar characteristics, resulting in e.g. aspecific binding in immunoassays. The microbial parameter can be linked with the method parameter **multifunctionality**.

Different pathogens or hygiene indicators that need to be detected or a variety of food types that need to be analyzed in the same lab all refer to multifunctional needs. With utility the total viable count is meant while hygiene indicators refer to microorganisms such as *Enterobacteriaceae* and *E. coli* (faecal) or *Staphylococcus aureus* (hands).

- **Managerial selection criteria**

The second main group of selection criteria are called the managerial selection criteria and is divided into i) the total time to obtain a result (TTD) and ii) the cost of the method. In general TTD, is either expressed as very rapid, i.e. result within 4-8 hours (same working day) or next day results (16-24 h) or multiple days (48-72 h), but faster then the "golden standard" reference method. TTD may differ for a negative test result, a presumptive test result and a confirmed positive test result. In a well working FSMS, many negative results are expected for pathogens, a rapid TTD for negative test results may be important. In monitoring studies by food business operators a presumptive positive test result may be important to start corrective actions although for recalls, alerts or legislative actions a confirmed positive result is generally awarded for and should be available as soon as possible.

The cost per sample is an important factor, but most of the time the cost per sample depends on the amount of samples or runs. The cost should be as low as possible. Consumables become cheaper if larger amounts are ordered at the same time. In addition the cost for investments and maintenance of equipment needs to be taken into account as well as the impact of personnel with a certain qualification or training. If a critical workload is attended automation may be considered. Automation often increases cost of investment whereas it reduces the input of personnel cost.

- **Technological selection criteria**

The technological selection criteria contains three main subcategories, i) the performance characteristic of the method (specificity, detection limit, etc.), ii) the operational requirements and III) infrastructure present. Overall, a satisfactory *performance* of the methods is taken for granted, and supported by the presence of a validation certificate, scientific publications and reports of method comparison studies by expert labs etc. Regarding the performance characteristic of quantitative methods the following method parameters are to be taken into account: **inclusivity, exclusivity, relative accuracy, detection limit, quantification limit, reproducibility, repeatability and linearity**. For the performance characteristic of detection methods the method parameters **specificity, sensitivity, relative accuracy and detection limit** are important. The ideal situation is a method with e.g.

specificity and sensitivity that equals 100%, but for a given test, sensitivity and specificity are often inversely related. Unfortunately not all this information is readily available for all methods or at least not for all food types in the scope. In the case of an independent validated method this information could be found in the validation report.

The *operational requirements* can be split up in two situations: the *routine analysis* and *ad hoc analysis*. The operational requirements are most of the time connected with each other. In contrast to ad hoc analysis, in routine analysis **high numbers of samples** need to be analyzed and an investment in specific **equipment** and or **automation** can be of interest. High-tech equipment, enabling automation and tracking and tracing, is mostly brought to the market with an extending product range, so also the parameter **multifunctional** is fulfilled. These equipments require an appropriate **training** for all the operators as well as for the designated maintenance team. In case of routine analysis the requirements convenience and user-friendly interface are highly valued whereas for an ad hoc analysis this will gain less weight.

The *infrastructure* is correlated with the required bio-safety level in the lab and the demands of lay out of working areas. In particular cases, there is a need for separate working areas in the lab or the availability of benches for dedicated equipment or manipulations to be done. Concerning the infrastructure to perform microbial analysis, two important "restrictions" can be noticed, one restriction regarding the **Bio-Safety Level** (BSL) and one regarding the **area classification**. For the detection of pathogens at least a Level 2 lab is needed while in case of manipulating *E. coli* O157 a higher level of bio-safety in the lab is needed, a Level 3 lab, although exact regulations may differ according to national regulations. Regarding area classification, for example for PCR purposes separated rooms are desired, one for the preparation of PCR mix (no DNA can be present), one where the DNA can be added to the PCR mix and one to run the PCR.

- **Sustainability selection criteria**

Sustainability criteria are consulted if more than one type of method can be taken under consideration for a MA after using the other criteria. Also in case of a defined type of method, but available in various formats from different suppliers this category may help out to finally select the best method fit for purpose in a particular situation. Under the term sustainability of the method, logistics becomes important and factors such as credibility of the supplier in terms of delivery, consumables, access to technical and administrative support, accessibility, availability of communication channels etc. are decisive in the final choice of method.

Most of the newly developed methods for microbial analysis aim for multifunctionality, rapid TTD and high throughput/automation and are thus rather directed to large capacity service labs or large companies having in-house labs. In the particular case of chromogenic agar and the lateral flow devices (LFD) only one sample can be analyzed per plate/LFD, but the advantages of this method are the simplicity in reading/interpretation of the result, fast (± 10 min) and the fact that no extra investment needs to be made regarding equipment. However it should be noticed that also the classical cultural methods are prone to evolve to respect the need for more rapid results e.g. enhancing culture media for rapid resuscitation and growth (e.g. new enrichment media) and inclusion of chromogenic agar plates.

Nowadays still in many labs the cultural methods based upon visualizing (phenotypical characteristics) and differentiating target microorganisms on agar plates (e.g. chromogenic media) are in use. Alternatively many labs use either immunological based (ELISA) or molecular based (PCR) methods which are also continuously evolving. As example of innovation, the total time to detect *L. monocytogenes* in foods by the yet already long established VIDAS® method is shortened due to the development of a new enrichment broth (*Listeria* eXpress broth) compared to the previous VIDAS® protocol. Another example is the combination of techniques: such as to combine IMS and PCR (e.g. Assurance GDS™ Biocontrol Systems)) using a convenient format of a pick-pen to concentrate bacteria and to eliminate PCR inhibiting compounds. Also the incorporation of new developments in available systems e.g. bacteriophage proteins instead of antibodies in ELFA (e.g. VIDAS®UP *E. coli* O157:H7 (BioMérieux)) is to be noticed as an upgrade of the system. Of course, these systems are rather directed to perform large capacity of samples. Most of the small and medium enterprises (SME's) in food industry outsource microbial analysis to service labs. Although, for their needs also developments are going on to provide on-the-spot-MA which are rapid, easy regarding interpretation and affordable. Especially for the enumeration of utility and hygiene indicators user-friendly methods are available. In the far future miniaturization such as lab-on-chip may also bring PCR techniques to the field.

7.4 Added value of this PhD and recommendation

- **Incorporation of sub-lethal injury into validation studies (ISO 16140)**

The current ISO 16140 method describes the following: “If it is not possible to acquire a sufficient number of naturally contaminated samples at the required levels, it is permissible to **contaminate artificially** some samples. The procedure of artificial contamination shall result in samples with analyte characteristics as similar to those in naturally contaminated samples. Therefore, it is written in Annex C of ISO 16140 in the case of spiked samples: a protocol for stressing the target microorganisms is to be defined and the stress demonstrated at the time of inoculation. **Standardized protocols** to induce sub-lethal injury are developed in **Chapter 2**. These types of protocols are recommended to spike samples according to the ISO 16140.

- **Potential false negative results for the presence of *E. coli* O157 using ISO 16654 or for the presence of *Campylobacter* spp. using ISO 10272-1**

The ISO methods were historically developed and decided based on international consensus, but holding on blind to these methods is shown to be dangerous. It was demonstrated in this PhD research that **potential risk** exist for **false negative results** using ISO 16654 methods (International Standards Organization, 2001a) for the detection of *E. coli* O157 from foods resulting in an underestimation of positive samples **(Chapter 4)**. The recommended enrichment broth mTSB+N has shown to be too selective if low numbers of sub-lethally injured cells of *E. coli* O157 are present, while it was observed that the non-selective BPW was able to resuscitate sub-lethally injured cells. Besides it was demonstrated that the use of Bolton broth (International Standards Organization, 1995) could result in an underestimation of true numbers of positive samples contaminated with *Campylobacter* spp. **(Chapter 5)** and highlighted the need to **revise the ISO method**. To correct for these underestimation Bayesian modelling can be applied (Habib *et al.*, 2008b), but continuous improvements are needed also for the ISO methods.

- **Rapid detection of the "big 4" food-borne pathogens**

From this PhD it became clear that regarding reliable detection of these food-borne pathogens the "18 h bottom" also referred to as the "overnight incubation" is difficult to break. Only for selected pathogens in selected foods, which means, in fact, food with small chance of the presence of sub-lethally injured cells, this "18 h bottom" can be improved. Alternative methods which claim to have come up with time to obtain results beneath "overnight procedures" e.g. around 12 h or less, ultimately fail when tested in more complex food matrices. It should be clear from this PhD that a universal "all foods" protocol is not realistic. In general, it can also be claimed that there is an inability to provide "overnight" results for all food matrices. For the moment there is still an "on-going" battle between technology providers to develop shorter time-to-result solutions, however despite all efforts, no major breakthrough has been accomplished the last few years. The known methodologies such as PCR, but also ELISA have an inherent floor which seems to be "reached" **(Chapter 6),** although techniques can be combined (such as IMS and PCR e.g. Assurance GDS™ Biocontrol Systems). Maybe "breakthrough" technologies, such as nanotechnology-based solutions, could yield better time to obtain results, while reaching same reliability level as conventional methods.

- **Selection criteria**

The selection tool helps the end user of the method to obtain a systematic insight into all relevant factors, beyond the technical performance characteristics of the method, to be taken into account for selection of a method for microbial analysis. It provides a rationale to facilitate the choice of an appropriate method in the frame of performance measurement of implemented food safety management systems (FSMS). It should be noticed that no method exist that is a 100% sensitive, and 100% specific, that can provide results without hands-on time and at low cost. All methods have advantages and disadvantages. The challenge is to select the best method for a particular purpose. Advantages of a method should be optimally exploited and the disadvantages should be recognized.

Although PCR is the leading technology in terms of time to obtain results, its use requires heavier investment, which may not be affordable for smaller laboratories or food business operators. However, also ELISA has been improved considerably and taken up the same solution for sample preparations as PCR also ending up in providing next day results (e.g. detection of *Salmonella* spp. with 3M™ Tecra™ Unique). As such the PCR method in speed has started to lose its competitive advantage that was first associated with the technique.

7.5 Perspectives for further research

Several reviewers last years focussed on sample preparation and detection of injured cells, showing that this bottleneck is clearly recognized (Benoit and Donahue, 2003; Brehm-Stecher *et al.*, 2009; Patel, 2000; Stevens and Jaykus, 2004). It is clear nowadays that sample preparation is the weakest link in the chain to provide fast and reliable results. Therefore further research is expected in the field of sample preparation in particular in developments of appropriate enrichment broths for several food-borne pathogens.

To enhance the development of new enrichment broths, more fundamental research is needed regarding sub-lethally injured cells. The effect of the newly used mild preservation techniques on different pathogenic bacteria should be examined (Gomez-Lopez *et al.*, 2009; Smigic *et al.*, 2009a). Molecular and/or microscopic techniques that can monitor the internal pH can be useful for this research (Shabala *et al.*, 2002; Siegumfeldt *et al.*, 1999; Siegumfeldt *et al.*, 2000; Smigic *et al.*, 2009b). Also microarrays, used in the framework of gene expression, are of importance for monitoring the effect of newly mild preservation techniques on bacteria. With the application of microarrays to study the expression of different genes, food microbiology is passing from the era of determining presence/absence or quantifying pathogenic microorganisms in foods to the era of determining the physiology of these microorganisms in different food-related environmental conditions and having the possibility to understand their virulence potential and behaviour (Abee *et al.*, 2004; Kato-Maeda *et al.*, 2001; Wells and Bennik, 2003). Understanding the behaviour of food-borne pathogens when present in foodstuffs can also contribute to the implementation of new mild preservation techniques to enhance food safety.

In this PhD dissertation many suggestions regarding enrichment broths have been made for different pathogens to be able to detect the possible sub-lethally injured cells present. In the future the robustness of these new method/enrichment protocols should be tested based on an interlaboratory study. Also more food matrices should be tested. Only after intensive/thoroughly research a method can be changed. In short the proposed new enrichment methods should be validated before changes can be made.

Different authors described already different growth promoting factors for the recovery of stressed *L. monocytogenes* cells (Besse, 2002; Beumer *et al.*, 1996; Busch and Donnelly, 1992; Chawla *et al.*, 1996; Dykes and Moorhead, 2001). Also a combination of different enrichment broths has shown its advantage in the recovery of *L. monocytogenes* from different matrices (Vlaemynck and Moermans, 1996). In the particular case of *Campylobacter* spp. detection, further investigation to formulate suitable enrichment broths or protocols is needed. Also a solution should be found for the problems raised in this PhD and investment/research should be done to obtain new enrichment broth/detection methods. Therefore more knowledge should be obtained towards the resuscitation and growth promoting factors in the broths for *Campylobacter* spp. Moreover, attention is to be given towards growth of possible interfering flora, especially under the influence of injury which may qualitative and quantitative differ between samples. Such enrichment broth should encompass all these differences and influences and include inter-strain variability. Moreover it needs to overcome consequences of intracellular differences within the population.

Further evolution of rapid detection methods is to be expected. Focus might be set on nanotechnology with devices such as biosensors (Maraldo and Mutharasan, 2007; Nordstrom *et al.*, 2008) and perhaps "lab on chip" in search of the wholly grail for online detection methods (Ruano-Lopez *et al.*, 2009).

Reference list

Reference list

1. Abee, T., van Schaik, W., Siezen, R.J., 2004. Impact of genomics on microbial food safety. Trends in Biotechnology 22, 653-660.

2. Adams, M. 2005. Detecting sublethally damaged cells. In: Griffiths, M. (Ed.), Understanding pathogen behaviour. Woodhead Publishing Limited. Cambridge, England, 199-212.

3. Aldus, C.F., van Amerongen, A., Ariens, R.M.C., Peck, M.W., Wichers, J.H., Wyatt, G.M., 2003. Principles of some novel rapid dipstick methods for detection and characterization of verotoxigenic *Escherichia coli*. Journal of Applied Microbiology 95, 380-389.

4. Altekruse, S.F., Stern, N.J., Fields, P.I., Swerdlow, D.L., 1999. *Campylobacter jejuni* - An emerging foodborne pathogen. Emerging Infectious Diseases 5, 28-35.

5. Amirlak, I., Amirlak, B., 2006. Haemolytic uraemic syndrome: An overview. Nephrology 11, 213-218.

6. Andrews, W.H., 1996. Validation of modern methods in food microbiology by AOAC INTERNATIONAL collaborative study. Food Control 7, 19-29.

7. Atlas, R.M, 2006. Handbook of microbiological media for the examination of food (2nd edition). CRC Press Taylor & Francis Group, Boca Raton, 1-446.

8. Baker, J.M., Griffiths, M.W., Collinsthompson, D.L., 1992. Bacterial Bioluminescence - Applications in Food Microbiology. Journal of Food Protection 55, 62-70.

9. Balamurugan, J., Bhilegaonkar, K.N., Agarwal, R.K., 2006. A study on suitability of four enrichment broths for PCR-based detection of *Listeria monocytogenes* from raw meat. Journal of Food Safety 26, 16-29.

10. Bautista, D.A., Chen, J., Barbut, S., Griffiths, M.W., 1998. Use of an autobioluminescent *Salmonella hadar* to monitor the effects of acid and temperature treatments on cell survival and viability on lactic acid-treated poultry carcasses. Journal of Food Protection 61, 1439-1445.

11. Bayles, D.O., Annous, B.A., Wilkinson, B.J., 1996. Cold stress proteins induced in *Listeria monocytogenes* in response to temperature downshock and growth at low temperatures. Applied and Environmental Microbiology 62, 1116-1119.

12. Baylis, C.L., MacPhee, S., Betts, R.P., 2000a. Comparison of two commercial preparations of buffered peptone water for the recovery and growth of *Salmonella* bacteria from foods. Journal of Applied Microbiology 89, 501-510.

13. Baylis, C.L., MacPhee, S., Martin, K.W., Humphrey, T.J., Betts, R.P., 2000b. Comparison of three enrichment media for the isolation of *Campylobacter* spp. from foods. Journal of Applied Microbiology 89, 884-891.

14. Begot, C., Desnier, I., Daudin, J.D., Labadie, J.C., Lebert, A., 1996. Recommendations for calculating growth parameters by optical density measurements. Journal of Microbiological Methods 25, 225-232.

15. Bell C. and Kyriakides A., 2002. *Salmonella*. A practical approach to the organism and its control in foods. Blackwell Science, London, 1-330.

16. Bell, B.P., Goldoft, M., Griffin, P.M., Davis, M.A., Gordon, D.C., Tarr, P.I., Bartleson, C.A., Lewis, J.H., Barrett, T.J., Wells, J.G., Baron, R., Kobayashi, J., 1994. A multistate outbreak of *Escherichia coli* O157:H7

associated bloody diarrhea and Hemolytic-Uremic-Syndrome from hamburgers - the Washington experience. Jama-Journal of the American Medical Association 272, 1349-1353.

17. Bell, C., Bowles, C.D., Toszeghy, M.J.K., Neaves, P., 1996. Development of a hygiene standard for raw milk based on the lumac ATP-bioluminescence method. International Dairy Journal 6, 709-713.

18. Beneduce, L., Spano, G., Massa, S., 2003. *Escherichia coli* O157:H7 general characteristics, isolation and identification techniques. Annals of Microbiology 53, 511-527.

19. Bennett, A.R., Davids, F.G.C., Vlahodimou, S., Banks, J.G., Betts, R.P., 1997. The use of bacteriophage-based systems for the separation and concentration of *Salmonella*. Journal of Applied Microbiology 83, 259-265.

20. Benoit, P.W., Donahue, D.W., 2003. Methods for rapid separation and concentration of bacteria in food that bypass time-consuming cultural enrichment. Journal of Food Protection 66, 1935-1948.

21. Berney, M., Vital, M., Hulshoff, I., Weilenmann, H.U., Egli, T., Hammes, F., 2008. Rapid, cultivation-independent assessment of microbial viability in drinking water. Water Research 42, 4010-4018.

22. Bertrand, S., Weill, F.X., Cloeckaert, A., Vrints, M., Mairiaux, E., Praud, K., Dierick, K., Wildemauve, C., Godard, C., Butaye, P., Imberechts, H., Grimont, P.A.D., Collard, J.M., 2006. Clonal emergence of extended-spectrum beta-lactamase (CTX-M-2)-producing *Salmonella enterica* serovar *Virchow* isolates with reduced susceptibilities to ciprofloxacin among poultry and humans in Belgium and France (2000 to 2003). Journal of Clinical Microbiology 44, 2897-2903.

23. Besse, N.G., 2002. Influence of various environmental parameters and of detection procedures on the recovery of stressed *L. monocytogenes*: a review. Food Microbiology 19, 221-234.

24. Besse, N.G., Audinet, N., Kerouanton, A., Colin, P., Kalmokoff, M., 2005. Evolution of *Listeria* populations in food samples undergoing enrichment culturing. International Journal of Food Microbiology 104, 123-134.

25. Besse, N.G., Brissonnet, F.D., Lafarge, V., Leclerc, V., 2000. Effect of various environmental parameters on the recovery of sublethally salt-damaged and acid-damaged *Listeria monocytogenes*. Journal of Applied Microbiology 89, 944-950.

26. Betts, R.P., Rentenaar, I.M.F., 1998. MicroVal: A European approach to the certification of new microbiological methods. Journal of Food Protection 61, 1579-1582.

27. Beumer, R.R., Hazeleger, W.C., 2003. *Listeria monocytogenes*: diagnostic problems. Fems Immunology and Medical Microbiology 35, 191-197.

28. Beumer, R.R., teGiffel, M.C., Anthonie, S.V.R., Cox, L.J., 1996. The effect of acriflavine and nalidixic acid on the growth of *Listeria* spp in enrichment media. Food Microbiology 13, 137-148.

29. Beutin, L., Jahn, S., Fach, P., 2009. Evaluation of the 'GeneDisc' real-time PCR system for detection of enterohaemorrhagic *Escherichia coli* (EHEC) O26, O103, O111, O145 and O157 strains according to their virulence markers and their O- and H-antigen-associated genes. Journal of Applied Microbiology 106, 1122-1132.

30. Bird, C.B., Miller, R.L., Miller, B.M., 1999. Reveal for *Salmonella* test system. Journal of Aoac International 82, 625-633.

31. Blackburn, C.D., McCarthy, J.D., 2000. Modifications to methods for the enumeration and detection of injured *Escherichia coli* O157:H7 in foods. International Journal of Food Microbiology 55, 285-290.

32. Blasco, R., Murphy, M.J., Sanders, M.F., Squirrell, D.J., 1998. Specific assays for bacteria using phage mediated release of adenylate kinase. Journal of Applied Microbiology 84, 661-666.

33. Blaser, M.J., Sazie, E., Williams, L.P., 1987. The influence of immunity on raw-milk-associated *Campylobacter* infection. Jama-Journal of the American Medical Association 257, 43-46.

34. Bohaychuk, V.M., Gensler, G.E., King, R.K., Wu, J.T., McMullen, L.M., 2005. Evaluation of detection methods for screening meat and poultry products for the presence of foodborne pathogens. Journal of Food Protection 68, 2637-2647.

35. Bolton, F.J., Crozier, L., Williamson, J.K., 1996. Isolation of *Escherichia coli* O157 from raw meat products. Letters in Applied Microbiology 23, 317-321.

36. Bottari, B., Ercolini, D., Gatti, M., Neviani, E., 2006. Application of FISH technology for microbiological analysis: current state and prospects. Applied Microbiology and Biotechnology 73, 485-494.

37. Botteldoorn, N., Van Coillie, E., Piessens, V., Rasschaert, G., Debruyne, L., Heyndrickx, M., Herman, L., Messens, W., 2008. Quantification of *Campylobacter* spp. in chicken carcass rinse by real-time PCR. Journal of Applied Microbiology 105, 1909-1918.

38. Bozoglu, F., Alpas, H., Kaletunc, G., 2004. Injury recovery of foodborne pathogens in high hydrostatic pressure treated milk during storage. Fems Immunology and Medical Microbiology 40, 243-247.

39. Brehm-Stecher, B., Young, C., Jaykus, L.A., Tortorello, M.L., 2009. Sample Preparation: The Forgotten Beginning. Journal of Food Protection 72, 1774-1789.

40. Bridson, E.Y., 2006. Oxoid the manual (9th edition). Oxoid Ltd, Basingstoke, 1-578.

41. Brouard, C., Espie, E., Weill, F.X., Kerouanton, A., Brisabois, A., Forgue, A.M., Vaillant, V., de Valk, H., 2007. Two consecutive large outbreaks of *Salmonella enterica* serotype *agona* infections in infants linked to the consumption of powdered infant formula. Pediatric Infectious Disease Journal 26, 148-152.

42. Busch, S.V., Donnelly, C.W., 1992. Development of a repair-enrichment broth for resuscitation of heat-injured *Listeria monocytogenes* and *Listeria innocua*. Applied and Environmental Microbiology 58, 14-20.

43. Busta, F.F., 1976. Practical implications of injured microorganisms in food. Journal of Milk and Food Technology 39, 138-145.

44. Call, D.R., Brockman, F.J., Chandler, D.P., 2001. Detecting and genotyping *Escherichia coli* O157:H7 using multiplexed PCR and nucleic acid microarrays. International Journal of Food Microbiology 67, 71-80.

45. Callicott, K.A., Haroardottir, H., Georgsson, F., Reiersen, J., Frioriksdottir, V., Gunnarsson, E., Michel, P., Bisaillon, J.R., Kristinsson, K.G., Briem, H., Hiett, K.L., Needleman, D.S., Stern, N.J., 2008. Broiler *Campylobacter* contamination and human campylobacteriosis in Iceland. Applied and Environmental Microbiology 74, 6483-6494.

46. Campbell, G.A., Delesdernier, D., Mutharasan, R., 2007a. Detection of airborne *Bacillus anthracis* spores by an integrated system of an air sampler and a cantilever immunosensor. Sensors and Actuators B-Chemical 127, 376-382.

47. Campbell, G.A., Medina, M.B., Mutharasan, R., 2007b. Detection of *Staphylococcus* enterotoxin B at picogram levels using piezoelectric-excited millimeter-sized cantilever sensors. Sensors and Actuators B-Chemical 126*,* 354-360.

48. Campbell, G.A., Mutharasan, R., 2005. Detection of pathogen *Escherichia coli* O157:H7 using self-excited PZT-glass microcantilevers. Biosensors & Bioelectronics 21*,* 462-473.

49. Campbell, G.A., Mutharasan, R., 2008. Near real-time detection of *Cryptosporidium parvum* oocyst by IgM-functionalized piezoelectric-excited millimeter-sized cantilever biosensor. Biosensors & Bioelectronics 23*,* 1039-1045.

50. Campbell, G.A., Uknalis, J., Tu, S.I., Mutharasan, R., 2007c. Detect of *Escherichia coli* O157:H7 in ground beef samples using piezoelectric excited millimeter-sized cantilever (PEMC) sensors. Biosensors & Bioelectronics 22*,* 1296-1302.

51. Canton, R., Novais, A., Valverde, A., Machado, E., Peixe, L., Baquero, F., Coque, T.M., 2008. Prevalence and spread of extended-spectrum beta-lactamase-producing *Enterobacteriaceae* in Europe. Clinical Microbiology and Infection 14*,* 144-153.

52. Cantoni, C., Giussani, A., Cocolin, L., 1998. Identification of *L. monocytogenes* and *Listeria* spp with a new ELISA sandwich. Industrie Alimentari 37*,* 176-177.

53. Carney, E., O'Brien, S.B., Sheridan, J.J., McDowell, D.A., Blair, I.S., Duffy, G., 2006. Prevalence and level of *Escherichia coli* O157 on beef trimmings, carcasses and boned head meat at a beef slaughter plant. Food Microbiology 23*,* 52-59.

54. Chandler, D.P., Brown, J., Call, D.R., Wunschel, S., Grate, J.W., Holman, D.A., Olson, L., Stottlemyre, M.S., Bruckner-Lea, C.J., 2001. Automated immunomagnetic separation and microarray detection of *E. coli* O157:H7 from poultry carcass rinse. International Journal of Food Microbiology 70, 143-154.

55. Chang, V.P., Mills, E.W., Cutter, C.N., 2003. Comparison of recovery methods for freeze-injured *Listeria monocytogenes*, *Salmonella typhimurium*, and *Campylobacter coli* in cell suspensions and associated with pork surfaces. Journal of Food Protection 66, 798-803.

56. Chapman, P.A., Malo, A.T.C., Siddons, C.A., Harkin, M., 1997. Use of commercial enzyme immunoassays and immunomagnetic separation systems for detecting *Escherichia coli* O157 in bovine fecal samples. Applied and Environmental Microbiology 63, 2549-2553.

57. Chawla, C.S., Chen, H., Donnelly, C.W., 1996. Mathematically modeling the repair of heat-injured *Listeria monocytogenes* as affected by temperature, pH, and salt concentration. International Journal of Food Microbiology 30, 231-242.

58. Chen, F.C., Godwin, S.L., 2006. Comparison of a rapid ATP bioluminescence assay and standard plate count methods for assessing microbial contamination of consumers' refrigerators. Journal of Food Protection 69, 2534-2538.

59. Chitarra, L.G., van den Bulk, R.W., 2003. The application of flow cytometry and fluorescent probe technology for detection and assessment of viability of plant pathogenic bacteria. European Journal of Plant Pathology 109, 407-417.

60. Chiu, C.H., Su, L.H., Chu, C., 2004. *Salmonella enterica* serotype *choleraesuis*: Epidemiology, pathogenesis, clinical disease, and treatment. Clinical Microbiology Reviews 17, 311-322.

61. Cohen, A.E., Kerdahi, K.F., 1996. Evaluation of a rapid and automated enzyme-linked fluorescent immunoassay say for detecting *Escherichia coli* serogroup O157 in cheese. Journal of Aoac International 79, 858-860.

62. Cone, R.W., Hobson, A.C., Huang, M.L.W., 1992. Coamplified positive control detects inhibition of polymerase chain reactions. Journal of Clinical Microbiology 30, 3185-3189.

63. Conner, D.E., Kotrola, J.S., 1995. Growth and survival of *Escherichia coli* O157:H7 under acidic conditions. Applied and Environmental Microbiology 61, 382-385.

64. Corbo, M.R., Bevilacqua, A., Campaniello, D., D'Amato, D., Speranza, B., Sinigaglia, M., 2009. Prolonging microbial shelf life of foods through the use of natural compounds and non-thermal approaches - a review. International Journal of Food Science and Technology 44, 223-241.

65. Cordier, J.L., 1994. HACCP in the chocolate industry. Food Control 5, 171-175.

66. Cornu, M., Kalmokoff, M., Flandrois, J.P., 2002. Modelling the competitive growth of *Listeria monocytogenes* and *Listeria innocua* in enrichment broths. International Journal of Food Microbiology 73, 261-274.

67. Corry, J.E.L., Post, D.E., Colin, P., Laisney, M.J., 1995. Culture Media for the Isolation of Campylobacters. International Journal of Food Microbiology 26, 43-76.

68. Cowart, R.E., Foster, B.G., 1985. Differential effects of iron on the growth of *Listeria monocytogenes* - minimum requirements and mechanism of acquisition. Journal of Infectious Diseases 151, 721-730.

69. Croci, L., Delibato, E., Volpe, G., Palleschi, G., 2001. A rapid electrochemical ELISA for the detection of salmonella in meat samples. Analytical Letters 34, 2597-2607.

70. Deisingh, A.K., Thompson, M., 2004. Strategies for the detection of *Escherichia coli* O157:H7 in foods. Journal of Applied Microbiology 96, 419-429.

71. Dostalek, P., Branyik, T., 2005. Prospects for rapid bioluminescent detection methods in the food industry - a review. Czech Journal of Food Sciences 23, 85-92.

72. Doyle, M.E., Mazzotta, A.S., Wang, T., Wiseman, D.W., Scott, V.N., 2001. Heat resistance of *Listeria monocytogenes*. Journal of Food Protection 64, 410-429.

73. Duffy, G., Cummins, E., Nally, P., O'Brien, S., Butler, F., 2006. A review of quantitative microbial risk assessment in the management of *Escherichia coli* O157:H7 on beef. Meat Science 74, 76-88.

74. Duffy, G., Garvey, P., 2000. Animal, food and biomedical aspects of verocytotoxigenic *E. coli*: an European concerted action project. Food Research International 33, 267-272.

75. Duffy, G., Sheridan, J.J., Buchanan, R.L., McDowell, D.A., Blair, I.S., 1994a. The effect of aeration, initial inoculum and meat microflora on the growth-kinetics of *Listeria monocytogenes* in selective enrichment broths. Food Microbiology 11, 429-438.

76. Duffy, L.L., Vanderlinde, P.B., Grau, F.H., 1994b. Growth of *Listeria monocytogenes* on vacuum-packed cooked meats - Effects of pH, Aw, nitrite and ascorbate. International Journal of Food Microbiology 23, 377-390.

77. Dykes, G.A., 1999. Physical and metabolic causes of sub-lethal damage in *Listeria monocytogenes* after long-term chilled storage at 4 degrees C. Journal of Applied Microbiology 87, 915-922.

78. Dykes, G.A., Moorhead, S.M., 2001. The role of L-carnitine and glycine betaine in the survival and sub-lethal injury of non-growing *Listeria monocytogenes* cells during chilled storage. Letters in Applied Microbiology 32, 282-286.

79. Dykes, G.A., Withers, K.M., 1999. Sub-lethal damage of *Listeria monocytogenes* after long-term chilled storage at 4 degrees C. Letters in Applied Microbiology 28, 45-48.

80. EFSA, 2009. The community summary report on trends and sources of zoonoses and zoonotic agents in the European Union in 2007. EFSA Journal, 1-217.

81. Eijkelkamp, J., Aarts, H., Fels-Klerx, H., 2009. Suitability of rapid detection methods for *Salmonella* in poultry slaughterhouses. Food Analytical Methods 2, 1-13.

82. Ercolini, D., 2004. PCR-DGGE fingerprinting: novel strategies for detection of microbes in food. Journal of Microbiological Methods 56, 297-314.

83. Erickson, M.C., Doyle, M.P., 2007. Food as a vehicle for transmission of Shiga toxin-producing *Escherichia coli*. Journal of Food Protection 70, 2426-2449.

84. EU, 2005. Commission regulation (EC) No 2073/2005 of 15 November 2005 on microbial criteria for foodstuffs. Official Journal of the European Union L338/1-26.

85. EUCAST, 2009. Breakpoint tables for interpretation of MICs and zone diameters, European Committee on Antimicrobial Susceptibility Testing.

Available at: http://www.eucast.org/fileadmin/src/media/PDFs/EUCAST_files/Disk_test_documents/EUCAST_breakpoints_v1.0_20091221.pdf.

86. Evrendilek, G.A., 2007. Survival of *Escherichia coli* O157:H7 in yogurt drink, plain yogurt and salted (tuzlu) yogurt: Effects of storage time, temperature, background flora and product characteristics. International Journal of Dairy Technology 60, 118-122.

87. Fakhr, M.K., Mcevoy, J.M., Sherwood, J.S., Logue, C.M., 2006. Adding a selective enrichment step to the iQ-Check (TM) real-time PCR improves the detection of *Salmonella* in naturally contaminated retail turkey meat products. Letters in Applied Microbiology 43, 78-83.

88. Farber, J.M., Peterkin, P.I., 1991. *Listeria monocytogenes*, a food-borne pathogen. Microbiological Reviews 55, 476-511.

89. Favrin, S.J., Jassim, S.A., Griffiths, M.W., 2003. Application of a novel immunomagnetic separation-bacteriophage assay for the detection of *Salmonella enteritidis* and *Escherichia coli* O157:H7 in food. International Journal of Food Microbiology 85, 63-71.

90. Feng, P., 1996. Emergence of rapid methods for identifying microbial pathogens in foods. Journal of Aoac International 79, 809-812.

91. Flanders, K.J., Donnelly, C.W., 1994. Injury, resuscitation and detection of *Listeria* spp. from frozen environments. Food Microbiology 11, 473-480.

92. Flanders, K.J., Pritchard, T.J., Donnelly, C.W., 1995. Enhanced recovery of *Listeria* from dairy-plant processing environments through combined use of repair enrichment and selective enrichment detection procedures. Journal of Food Protection 58, 404-409.

93. Flint, S., Drocourt, J.L., Walker, K., Stevenson, B., Dwyer, M., Clarke, I., McGill, D., 2006. A rapid, two-hour method for the enumeration of total viable bacteria in samples from commercial milk powder and whey protein concentrate powder manufacturing plants. International Dairy Journal 16, 379-384.

94. Flint, S., Walker, K., Waters, B., Crawford, R., 2007. Description and validation of a rapid (1h) flow cytometry test for enumerating thermophilic bacteria in milk powders. Journal of Applied Microbiology 102, 909-915.

95. Fluit, A.C., Widjojoatmodjo, M.N., Box, A.T.A., Torensma, R., Verhoef, J., 1993. Rapid detection of *Salmonellae* in poultry with the magnetic immuno-polymerase chain reaction assay. Applied and Environmental Microbiology 59, 1342-1346.

96. Francois, K., Devlieghere, F., Uyttendaele, M., Standaert, A.R., Geeraerd, A.H., Nadal, P., Van Impe, J.F., Debevere, J., 2006. Single cell variability of *L. monocytogenes* grown on liver pate and cooked ham at 7°C: comparing challenge test data to predictive simulations. Journal of Applied Microbiology 100, 800-812.

97. Fredriksson-Ahomaa, M., Stolle, A., Siitonen, A., Korkeala, H., 2006. Sporadic human Yersinia enterocolitica infections caused by bioserotype 4/O:3 originate mainly from pigs. Journal of Medical Microbiology, 55, 747-749.

98. Fukuda, S., Tatsumi, H., Igimi, S., Yamamoto, S., 2005. Improved bioluminescent enzyme immunoassay for the rapid detection of *Salmonella* in chicken meat samples. Letters in Applied Microbiology 41, 379-384.

99. Fukushima, H., Katsube, K., Hata, Y., Kishi, R., Fujiwara, S., 2007. Rapid separation and concentration of food-borne pathogens in food samples

prior to quantification by viable-cell counting and real-time PCR. Applied and Environmental Microbiology 73, 92-100.

100. Fung, D.Y.C., 1994. Rapid methods and automation in food microbiology - A review. Food Reviews International 10, 357-375.

101. Gantois, I., Ducatelle, R., Pasmans, F., Haesebrouck, F., Gast, R., Humphrey, T.J., Van Immerseel, F., 2009. Mechanisms of egg contamination by *Salmonella* Enteritidis. Fems Microbiology Reviews 33, 718-738.

102. Gomez-Lopez, V.M., Rajkovic, A., Ragaert, P., Smigic, N., Devlieghere, F., 2009. Chlorine dioxide for minimally processed produce preservation: a review. Trends in Food Science & Technology 20, 17-26.

103. Gooding, C.M., Choudary, P.V., 1998. Detection of *Escherichia coli* O157:H7 in ground beef in eight hours. Journal of Microbiological Methods 34, 89-98.

104. Goodridge, L., Griffiths, M., 2002. Reporter bacteriophage assays as a means to detect foodborne pathogenic bacteria. Food Research International 35, 863-870.

105. Gracias, K.S., McKillip, J.L., 2004. A review of conventional detection and enumeration methods for pathogenic bacteria in food. Canadian Journal of Microbiology 50, 883-890.

106. Graham, S.M., 2002. Salmonellosis in children in developing and developed countries and populations. Current Opinion in Infectious Diseases 15, 507-512.

107. Grant, M.A., 2004. Improved laboratory enrichment for enterohemorrhagic *Escherichia coli* by exposure to extremely acidic conditions. Applied and Environmental Microbiology 70, 1226-1230.

108. Grant, M.A., 2005. Comparison of a new enrichment procedure for Shiga toxin-producing *Escherichia coli* with five standard methods. Journal of Food Protection 68, 1593-1599.

109. Grant, M.A., 2008. Comparison of *Escherichia coli* O157:H7 enrichment in spiked produce samples. Journal of Food Protection 71, 139-145.

110. Gray, K.M., Bhunia, A.K., 2005. Specific detection of cytopathogenic *Listeria monocytogenes* using a two-step method of immunoseparation and cytotoxicity analysis. Journal of Microbiological Methods 60, 259-268.

111. Greenland K., de Jager C., Heuvelink A., van der Zwaluw K., Heck M., Notermans D., van Pelt W., Friesema I., 2009. Nationwide outbreak of STEC O157 infection in the Netherlands, December 2008-January 2009: continuous risk of consuming raw beef products. Euro Surveillance 14, pii=19129. Available at : http://www.eurosurveillance.org/ViewArticle.aspx?ArticleId=19129.

112. Greenwood, M., Willis, C., Doswell, P., Allen, G., Pathak, K., 2005. Evaluation of chromogenic media for the detection of *Listeria* species in food. Journal of Applied Microbiology 99, 1340-1345.

113. Gryson, N., Dewettinck, K., Messens, K., 2007. Influence of cocoa components on the PCR detection of soy lecithin DNA. European Food Research and Technology 226, 247-254.

114. Gryson, N., Messens, K., Dewettinck, K., 2004. Evaluation and optimisation of five different extraction methods for soy DNA in chocolate and biscuits. Extraction of DNA as a first step in GMO analysis. Journal of the Science of Food and Agriculture 84, 1357-1363.

115. Gunasekera, T.S., Veal, D.A., Attfield, P.V., 2003. Potential for broad applications of flow cytometry and fluorescence techniques in

microbiological and somatic cell analyses of milk. International Journal of Food Microbiology 85, 269-279.

116. Habib, I., Sampers, I., Uyttendaele, M., Berkvens, D., De Zutter, L., 2008a. Baseline data from a Belgium-wide survey of *Campylobacter* species contamination in chicken meat preparations and considerations for a reliable monitoring program. Applied and Environmental Microbiology 74, 5483-5489.

117. Habib, I., Sampers, I., Uyttendaele, M., De Zutter, L., Berkvens, D., 2008b. A Bayesian modelling framework to estimate *Campylobacter* prevalence and culture methods sensitivity: application to a chicken meat survey in Belgium. Journal of Applied Microbiology 105, 2002-2008.

118. Hagens, S., Loessner, M.J., 2007. Application of bacteriophages for detection and control of foodborne pathogens. Applied Microbiology and Biotechnology 76, 513-519.

119. Hagens, S., Offerhaus, M.L., 2008. Bacteriophages - New weapons for food safety. Food Technology 62, 46-54.

120. Hanna, S.E., Connor, C.J., Wang, H.H., 2005. Real-time polymerase chain reaction for the food microbiologist: Technologies, applications, and limitations. Journal of Food Science 70, R49-R53.

121. Hara-Kudo, Y., Ikedo, M., Kodaka, H., Nakagawa, H., Goto, K., Masuda, T., Konuma, H., Kojima, T., Kumagai, S., 2000. Selective enrichment with a resuscitation step for isolation of freeze-injured *Escherichia coli* O157:H7 from foods. Applied and Environmental Microbiology 66, 2866-2872.

122. Harris, L.J., Griffiths, M.W., 1992. The detection of foodborne pathogens by the polymerase chain reaction (PCR). Food Research International 25, 457-469.

123. Hazeleger, W.C., Beumer, R.R., Rombouts, F.M., 1992. The use of latex agglutination tests for determining *Campylobacter* species. Letters in Applied Microbiology 14, 181-184.

124. Hendrickx, M., Maesmans, G., Decordt, S., Noronha, J., Vanloey, A., Tobback, P., 1995. Evaluation of the integrated time-temperature effect in thermal-processing of foods. Critical Reviews in Food Science and Nutrition 35, 231-262.

125. Hepburn, N.F., MacRae, M., Johnston, M., Mooney, J., Ogden, I.D., 2002. Optimizing enrichment conditions for the isolation of *Escherichia coli* O157 in soils by immunomagnetic separation. Letters in Applied Microbiology 34, 365-369.

126. Herman, L.M.F., Deblock, J.H.G.E., Moermans, R.J.B., 1995. Direct detection of *Listeria monocytogenes* in 25 milliliters of raw-milk by a 2-step PCR with nested primers. Applied and Environmental Microbiology 61, 817-819.

127. Hong, B.X., Jiang, L.F., Hu, Y.S., Fang, D.Y., Guo, H.Y., 2004. Application of oligonucleotide array technology for the rapid detection of pathogenic bacteria of foodborne infections. Journal of Microbiological Methods 58, 403-411.

128. Hoorfar, J., Cook, N., Malorny, B., Wagner, M., De Medici, D., Abdulmaajood, A., Fach, P., 2003. Making internal amplification control mandatory for diagnostic PCR. Journal of Clinical Microbiology 41, 5835.

129. Hoorfar, J., Cook, N., Malorny, B., Wagner, M., De Medici, D., Abdulmawjood, A., Fach, P., 2004. Diagnostic PCR: Making internal amplification control mandatory. Journal of Applied Microbiology 96, 221-222.

130. Humphrey, T., O'Brien, S., Madsen, M., 2007. Campylobacters as zoonotic pathogens: A food production perspective. International Journal of Food Microbiology 117, 237-257.

131. Hurst, A., 1977. Bacterial injury - Review. Canadian Journal of Microbiology 23, 935-944.

132. Hussein, H.S., Bollinger, L.M., Hall, M.R., 2008. Growth and enrichment medium for detection and isolation of Shiga toxin-producing *Escherichia coli* in cattle feces. Journal of Food Protection 71, 927-933.

133. Ikeda, M., Yamaguchi, N., Tani, K., Nasu, M., 2006. Detection of food poisoning bacteria in fresh vegetables using DNA microarray. Journal of Health Science 52, 36-42.

134. In 't Veld, P.H., 1998. The use of reference materials in quality assurance programmes in food microbiology laboratories. International Journal of Food Microbiology 45, 35-41.

135. In 't Veld, P.H., Notermans, S., 1992. Use of Reference Materials (Spray-dried milk artificially contaminated with *Salmonella* Typhimurium) to validate detection methods for *Salmonella*. Journal of Food Protection 55, 855-858.

136. In 't Veld, P.H., Soentoro, P.S.S., Notermans, S.H.W., 1993. Properties of *Bacillus cereus* spores in reference materials prepared from artificially contaminated spray-dried milk. International Journal of Food Microbiology 20, 23-36.

137. International Standards Organization, 1995. ISO 10272: Microbiology of food and animal feeding stuffs -- Horizontal method for detection and enumeration of *Campylobacter* spp. -- Part 1: Detection method.

138. International Standards Organization, 1996. ISO 11290: Microbiology of food and animal feeding stuffs -- Horizontal method for the detection and enumeration of *Listeria monocytogenes*.

139. International Standards Organization, 2001a. ISO 16654: Microbiology of food and animal feeding stuffs -- Horizontal method for the detection of *Escherichia coli* O157.

140. International Standards Organization, 2001b. ISO 16649: Microbiology of food and animal feeding stuffs -- Horizontal method for the enumeration of beta-glucuronidase-positive *Escherichia coli*.

141. International Standards Organization, 2003a. ISO 4833: Microbiology of food and animal feeding stuffs -- Horizontal method for the enumeration of microorganisms -- Colony-count technique at 30 degrees C.

142. International Standards Organization, 2003b. ISO 16140: Microbiology of food and animal feeding stuff -- Protocol for the validation of alternative methods.

143. International Standards Organization, 2004. ISO 11290-2:1998 FDAM:2004(E): Microbiology of food and animal feeding stuff – Horizontal method for the detection and enumeration of *Listeria monocytogenes*— Part2: Enumeration method. Amendment 1: Modification of the enumeration medium.

144. International Standards Organization, 2005a. ISO 17025: General requirements for the competence of testing and calibration laboratories.

145. International Standards Organization, 2005b. ISO 6579: Microbiology of food and animal feeding stuffs -- Horizontal method for the detection of *Salmonella* spp.

146. International Standard Organization, 2006a. ISO 19036: Microbiology of food and animal feeding stuffs -- Guidelines for the estimation of measurement uncertainty for quantitative determinations.

147. International Standards Organization, 2006b. ISO 10272: Microbiology of food and animal feeding stuffs -- Horizontal method for detection and enumeration of *Campylobacter* spp. -- Part 1: Detection method.

148. Ivnitski, D., Abdel-Hamid, I., Atanasov, P., Wilkins, E., 1999. Biosensors for detection of pathogenic bacteria. Biosensors & Bioelectronics 14, 599-624.

149. Jacxsens, L., Kussaga, J., Luning, P.A., Van der Spiegel, M., Uyttendaele, M., 2009. A microbial assessment scheme to support microbial performance measurements of food safety management systems. International Journal of Food Microbiology 134, 113-125.

150. Janning, B., Tveld, P.H., Mooijman, K.A., Havelaar, A.H., 1995. Development, production and certification of microbiological reference materials. Fresenius Journal of Analytical Chemistry 352, 240-245.

151. Jantzen, M.M., Navas, J., de Paz, M., Rodriguez, B., da Silva, W.P., Nunez, M., Martinez-Suarez, J.V., 2006. Evaluation of ALOA plating medium for its suitability to recover high pressure-injured *Listeria monocytogenes* from ground chicken meat. Letters in Applied Microbiology 43, 313-317.

152. Jaykus, L.A., 2003. Challenges to developing real-time methods to detect pathogens in foods. Asm News 69, 341-347.

153. Jeffries, L., 1959. Novobiocin-Tetrathionate Broth - A medium of improved selectivity for the isolation of *Salmonellae* from faeces. Journal of Clinical Pathology 12, 568-571.

154. Johansson, T., 1998. Enhanced detection and enumeration of *Listeria monocytogenes* from foodstuffs and food-processing environments. International Journal of Food Microbiology 40, 77-85.

155. Jordan, K.N., Davies, K.W., 2001. Sodium chloride enhances recovery and growth of acid-stressed *E. coli* O157:H7. Letters in Applied Microbiology 32, 312-315.

156. Joux, F., Lebaron, P., 2000. Use of fluorescent probes to assess physiological functions of bacteria at single-cell level. Microbes and Infection 2, 1523-1535.

157. Kamioka, I., Nozu, K., Fujita, T., Kaito, H., Tanaka, R., Yoshiya, K., Iijima, K., Nakanishi, K., Yoshikawa, N., Matsuo, M., 2007. Prognosis and pathological characteristics of five children with non-Shiga toxin-mediated hemolytic uremic syndrome. Pediatrics International 49, 196-201.

158. Karch, H., Tarr, P.I., Bielaszewska, M, 2005. Enterohaemorrhagic *Escherichia coli* in human medicine. International Journal of Medical Microbiology 295, 404-418.

159. Kato-Maeda, M., Gao, Q., Small, P.M., 2001. Microarray analysis of pathogens and their interaction with hosts. Cellular Microbiology 3, 713-719.

160. Kawasaki, S., Fratamico, P.M., Horikoshi, N., Okada, Y., Takeshita, K., Sameshima, T., Kawamoto, S., 2009. Evaluation of a multiplex PCR system for simultaneous detection of *Salmonella* spp., *Listeria monocytogenes*, and *Escherichia coli* O157:H7 in foods and in food subjected to freezing. Foodborne Pathogens and Disease 6, 81-89.

161. Keith, M., 1997. Evaluation of an automated enzyme-linked fluorescent immunoassay system for the detection of salmonellae in foods. Journal of Food Protection 60, 682-685.

162. Kell, D.B., Kaprelyants, A.S., Weichart, D.H., Harwood, C.R., Barer, M.R., 1998. Viability and activity in readily culturable bacteria: a review and discussion of the practical issues. Antonie Van Leeuwenhoek International Journal of General and Molecular Microbiology 73, 169-187.

163. Keramas, G., Bang, D.D., Lund, M., Madsen, M., Rasmussen, S.E., Bunkenborg, H., Telleman, P., Christensen, C.B.V., 2003. Development of a sensitive DNA microarray suitable for rapid detection of *Campylobacter* spp. Molecular and Cellular Probes 17, 187-196.

164. Komitopoulou, E., Penaloza, W., 2009. Fate of *Salmonella* in dry confectionery raw materials. Journal of Applied Microbiology 106, 1892-1900.

165. Kostrzynska, M., Bachand, A., 2006. Application of DNA microarray technology for detection, identification, and characterization of food-borne pathogens. Canadian Journal of Microbiology 52, 1-8.

166. Kothary, M.H., Babu, U.S., 2001. Infective dose of foodborne pathogens in volunteers: A review. Journal of Food Safety 21, 49-73.

167. Kotzekidou, P., 1998. Microbial stability and fate of *Salmonella enteritidis* in halva, a low-moisture confection. Journal of Food Protection 61, 181-185.

168. Kozak, J., Balmer, T., Byrne, R., Fisher, K., 1996. Prevalence of *Listeria monocytogenes* in foods: Incidence in dairy products. Food Control 7, 215-221.

169. Krascsenicsova, K., Piknova, L., Kaclikova, E., Kuchta, T., 2008. Detection of *Salmonella enterica* in food using two-step enrichment and real-time polymerase chain reaction. Letters in Applied Microbiology 46, 483-487.

170. Krause, M., Josefsen, M.H., Lund, M., Jacobsen, N.R., Brorsen, L., Moos, M., Stockmarr, A., Hoorfar, J., 2006. Comparative, collaborative, and on-site validation of a TaqMan PCR method as a tool for certified production of fresh, *Campylobacter*-free chickens. Applied and Environmental Microbiology 72*,* 5463-5468.

171. Kretzer, J.W., Lehmann, R., Schmelcher, M., Banz, M., Kim, K.P., Korn, C., Loessner, M.J., 2007. Use of high-affinity cell wall-binding domains of bacteriophage endolysins for immobilization and separation of bacterial cells. Applied and Environmental Microbiology 73*,* 1992-2000.

172. Kumar, R., Surendran, P.K., Thampuran, N., 2008. Evaluation of culture, ELISA and PCR assays for the detection of *Salmonella* in seafood. Letters in Applied Microbiology 46*,* 221-226.

173. Kurbanoglu, E.B., Algur, O.F., 2006. Utilization of ram horn hydrolysate as a supplement for recovery of heat- and freeze-injured bacteria. Food Control 17*,* 238-242.

174. Lachica, R.V., 1990. Simplified Henry technique for initial recognition of *Listeria* colonies. Applied and Environmental Microbiology 56*,* 1164-1165.

175. Law, D., 2000. Virulence factors of *Escherichia coli* O157 and other Shiga toxin-producing *E. coli*. Journal of Applied Microbiology 88*,* 729-745.

176. Lazcka, O., Del Campo, F.J., Munoz, F.X., 2007. Pathogen detection: A perspective of traditional methods and biosensors. Biosensors & Bioelectronics 22*,* 1205-1217.

177. LeBlanc, J.J., 2003. Implication of virulence factors in *Escherichia coli* O157:H7 pathogenesis. Critical Reviews in Microbiology 29*,* 277-296.

178. Leistner, L., Gorris, L.G.M., 1995. Food preservation by Hurdle technology. Trends in Food Science & Technology 6*,* 41-46.

179. Levin, R.E., 2007. *Campylobacter jejuni*: A review of its characteristics, pathogenicity, ecology, distribution, subspecies characterization and molecular methods of detection. Food Biotechnology 21, 271-347.

180. Leyer, G.J., Wang, L.L., Johnson, E.A., 1995. Acid adaptation of *Escherichia coli* O157:H7 increases survival in acidic foods. Applied and Environmental Microbiology 61, 3752-3755.

181. Li, Y.Q., Odumeru, J.A., Griffiths, M., McKellar, R.C., 2006. Effect of environmental stresses on the mean and distribution of individual cell lag times of Escherichia coli O157:H7. International Journal of Food Microbiology 110, 278-285.

182. Lianou, A., Sofos, J.N., 2007. A review of the incidence and transmission of *Listeria monocytogenes* in ready-to-eat products in retail and food service environments. Journal of Food Protection 70, 2172-2198.

183. Liao, C.H., Fett, W.F., 2005. Resuscitation of acid-injured *Salmonella* in enrichment broth, in apple juice and on the surfaces of fresh-cut cucumber and apple. Letters in Applied Microbiology 41, 487-492.

184. Lilja, L., Hanninen, M.L., 2001. Evaluation of a commercial automated ELISA and PCR-method for rapid detection and identification of *Campylobacter jejuni* and *C. coli* in poultry products. Food Microbiology 18, 205-209.

185. Lindqvist, R., 1997. Preparation of PCR samples from food by a rapid and simple centrifugation technique evaluated by detection of *Escherichia coli* O157:H7. International Journal of Food Microbiology 37, 73-82.

186. Lindqvist, R., Norling, B., Lambertz, S.T., 1997. A rapid sample preparation method for PCR detection of food pathogens based on

buoyant density centrifugation. Letters in Applied Microbiology 24, 306-310.

187. Lionberg, W.C., Restaino, L., Frampton, E.W., Barbour, W.M., 2003. Efficacy of enrichment broths in the recovery of freeze-injured *Escherichia coli* O157:H7 in inoculated ground beef by PCR. Journal of Food Protection 66, 1911-1915.

188. Little, C.L., Gillespie, I.A., 2008. Prepared salads and public health. Journal of Applied Microbiology 105, 1729-1743.

189. Livermore, D.M., Hope, R., Brick, G., Lillie, M., Reynolds, R., 2008. Non-susceptibility trends among *Enterobacteriaceae* from bacteraemias in the UK and Ireland, 2001-06. Journal of Antimicrobial Chemotherapy 62, ii41-ii54.

190. Loessner, M.J., Kramer, K., Ebel, F., Scherer, S., 2002. C-terminal domains of *Listeria monocytogenes* bacteriophage murein hydrolases determine specific recognition and high-affinity binding to bacterial cell wall carbohydrates. Molecular Microbiology 44, 335-349.

191. Lund, M., Nordentoft, S., Pedersen, K., Madsen, M., 2004. Detection of *Campylobacter* spp. in chicken fecal samples by real-time PCR. Journal of Clinical Microbiology 42, 5125-5132.

192. Luo, P.G., Stutzenberger, F.J., 2008. Nanotechnology in the detection and control of microorganisms. Advances in Applied Microbiology 63, 145-181.

193. Machado, E., Coque, T.M., Canton, R., Sousa, J.C., Peixe, L., 2008. Antibiotic resistance integrons and extended-spectrum beta-lactamases among *Enterobacteriaceae* isolates recovered from chickens and swine in Portugal. Journal of Antimicrobial Chemotherapy 62, 296-302.

194. Mackey, B.M., Boogard, E., Hayes, C.M., Baranyi, J., 1994. Recovery of heat-injured *Listeria monocytogenes.* International Journal of Food Microbiology 22, 227-237.

195. Malorny, B., Hoorfar, J., Hugas, M., Heuvelink, A., Fach, P., Ellerbroek, L., Bunge, C., Dorn, C., Helmuth, R., 2003a. Interlaboratory diagnostic accuracy of a *Salmonella* specific PCR-based method. International Journal of Food Microbiology 89, 241-249.

196. Malorny, B., Tassios, P.T., Radstrom, P., Cook, N., Wagner, M., Hoorfar, J., 2003b. Standardization of diagnostic PCR for the detection of foodborne pathogens. International Journal of Food Microbiology 83, 39-48.

197. Manafi, M., 1996. Fluorogenic and chromogenic enzyme substrates in culture media and identification tests. International Journal of Food Microbiology 31, 45-58.

198. Manafi, M., 2000. New developments in chromogenic and fluorogenic culture media. International Journal of Food Microbiology 60, 205-218.

199. Maraldo, D., Mutharasan, R., 2007. 10-minute assay for detecting *Escherichia coli* O157:H7 in ground beef samples using piezoelectric-excited millimeter-size cantilever sensors. Journal of Food Protection 70, 1670-1677.

200. Marcus, R., 2008. New information about pediatric foodborne infections: the view from FoodNet. Current Opinion in Pediatrics 20, 79-84.

201. Matthews, L., McKendrick, I.J., Ternent, H., Gunn, G.J., Synge, B., Woolhouse, M.E.J., 2006. Super-shedding cattle and the transmission dynamics of *Escherichia coli* O157. Epidemiology and Infection 134, 131-142.

202. Maurer J., 2006. PCR methods in food. Springer Science+Business Media Inc., New York, 1-148.

203. McClure, P.J., Hall, S., 2000. Survival of *Escherichia coli* in foods. Journal of Applied Microbiology 88, 61-70.

204. McKellar, R.C., 1994. Use of the CAMP Test for identification of *Listeria monocytogenes*. Applied and Environmental Microbiology 60, 4219-4225.

205. McLauchlin, J., Mitchell, R.T., Smerdon, W.J., Jewell, K., 2004. *Listeria monocytogenes* and listeriosis: a review of hazard characterisation for use in microbiological risk assessment of foods. International Journal of Food Microbiology 92, 15-33.

206. McMahon, W.A., Schultz, A.M., Johnson, R.L., 2004. Evaluation of VIDAS((R)) *Salmonella* (SLM) immunoassay method with Rappaport-Vassiliadis (RV) medium for detection of *Salmonella* in foods: Collaborative study. Journal of Aoac International 87, 867-883.

207. Meighen, E.A., 1991. Molecular-Biology of Bioluminescence. Bioluminescence and Chemiluminescence 3-10.

208. Meighen, E.A., 1993. Bacterial bioluminescence - organization, regulation, and application of the *lux* genes. Faseb Journal 7, 1016-1022.

209. Mellefont, L.A., McMeekin, T.A., Ross, T., 2005. Viable count estimates of lag time responses for *Salmonella typhimurium* M48 subjected to abrupt osmotic shifts. International Journal of Food Microbiology 105, 399-410.

210. Miller, L.G., Kaspar, C.W., 1994. *Escherichia coli* O157:H7 acid tolerance and survival in apple cider. Journal of Food Protection 57, 460-464.

211. Miller, R.S., Speegle, L., Oyarzabal, O.A., Lastovica, A.J., 2008. Evaluation of three commercial latex agglutination tests for identification of *Campylobacter* spp. Journal of Clinical Microbiology 46, 3546-3547.

212. Molina, P.M., Parma, A.E., Sanz, M.E., 2003. Survival in acidic and alcoholic medium of Shiga toxin-producing *Escherichia coli* O157:H7 and non-O157:H7 isolated in Argentina. BMC Microbiology 3, 1-6.

213. Moorhead, S.M., Dykes, G.A., 2002. Survival of *Campylobacter jejuni* on beef trimmings during freezing and frozen storage. Letters in Applied Microbiology 34, 72-76.

214. Moter, A., Gobel, U.B., 2000. Fluorescence in situ hybridization (FISH) for direct visualization of microorganisms. Journal of Microbiological Methods 41, 85-112.

215. Mullis, K.B., Faloona, F.A., 1987. Specific synthesis of DNA invitro via a polymerase-catalyzed chain reaction. Methods in Enzymology 155, 335-350.

216. Murphy, C., Carroll, C., Jordan, K.N., 2003. Induction of an adaptive tolerance response in the foodborne pathogen, *Campylobacter jejuni*. Fems Microbiology Letters 223, 89-93.

217. Murphy, C., Carroll, C., Jordan, K.N., 2005. The effect of different media on the survival and induction of stress responses by *Campylobacter jejuni*. Journal of Microbiological Methods 62, 161-166.

218. Murphy, C., Carroll, C., Jordan, K.N., 2006. Environmental survival mechanisms of the foodborne pathogen *Campylobacter jejuni*. Journal of Applied Microbiology 100, 623-632.

219. Musgrove, M.T., Berrang, M.E., Byrd, J.A., Stern, N.J., Cox, N.A., 2001. Detection of *Campylobacter* spp. in ceca and crops with and without enrichment. Poultry Science 80, 825-828.

220. Muthukumarasamy, P., Holley, R.A., 2007. Survival of *Escherichia coli* O157:H7 in dry fermented sausages containing micro-encapsulated probiotic lactic acid bacteria. Food Microbiology 24, 82-88.

221. Nachamkin, I., Szymanski, C.M. and Blaser M.J., 2008. *Campylobacter* (3th Edition). ASM Press, Washington DC, 1-732.

222. Ngutter, C., Donnelly, C., 2003. Nitrite-induced injury of *Listeria monocytogenes* and the effect of selective versus nonselective recovery procedures on its isolation from frankfurters. Journal of Food Protection 66, 2252-2257.

223. Niskanen, A., Korkeala, H., Manninen, M., Vuento, M., Kuusela, P., 1991. Evaluation of 3 slide agglutination tests for rapid identification of *Staphylococcus aureus*. Acta Veterinaria Scandinavica 32, 543-549.

224. Niza-Ribeiro, J., Louza, A.C., Santos, P., Lima, M., 2000. Monitoring the microbiological quality of raw milk through the use of an ATP bioluminescence method. Food Control 11, 209-216.

225. Nogva, H.K., Rudi, K., Naterstad, K., Holck, A., Lillehaug, D., 2000. Application of 5 '-nuclease PCR for quantitative detection of *Listeria monocytogenes* in pure cultures, water, skim milk, and unpasteurized whole milk. Applied and Environmental Microbiology 66, 4266-4271.

226. Nordstrom, M., Keller, S., Lillemose, M., Johansson, A., Dohn, S., Haefliger, D., Blagoi, G., Havsteen-Jakobsen, M., Boisen, A., 2008. SU-8 cantilevers for bio/chemical sensing; Fabrication, characterisation and development of novel read-out methods. Sensors 8, 1595-1612.

227. Novotny, L., Dvorska, L., Lorencova, A., Beran, V., Pavlik, I., 2004. Fish: a potential source of bacterial pathogens for human beings. Veterinarni Medicina 49*,* 343-358.

228. Ogawa, R., Pacelli, R., Espey, M.G., Miranda, K.M., Friedman, N., Kim, S.M., Cox, G., Mitchell, J.B., Wink, D.A., Russo, A., 2001. Comparison of control of *Listeria* by nitric oxide redox chemistry from murine macrophages and no donors: Insights into listeriocidal activity of oxidative and nitrosative stress. Free Radical Biology and Medicine 30*,* 268-276.

229. Okrend, A.J.G., Rose, B.E., Lattuada, C.P., 1992. Isolation of *Escherichia coli* O157:H7 using O157 specific antibody coated magnetic beads. Journal of Food Protection 55*,* 214-217.

230. Oksuz, O., Arici, M., Kurultay, S., Gumus, T., 2004. Incidence of *Escherichia coli* O157 in raw milk and white pickled cheese manufactured from raw milk in Turkey. Food Control 15*,* 453-456.

231. Osborne, C.M., Bremer, P.J., 2002. Development of a technique to quantify the effectiveness of enrichment regimes in recovering "stressed" *Listeria* cells. Journal of Food Protection 65*,* 1122-1128.

232. Osmanagaoglu, O., 2005. Sensitivity of sublethally injured Gram-negative bacteria to pediocin P. Journal of Food Safety 25*,* 266-275.

233. Patel, J.R., Bhagwat, A.A., 2008. Rapid real-time PCR assay for detecting *Salmonella* in raw and ready-to-eat meats. Acta Veterinaria Hungarica 56*,* 451-458.

234. Patel, P., 2000. A review of analytical, separation, concentration and segregation techniques in microbiology. Journal of Rapid Methods and Automation in Microbiology 8*,* 227-248.

235. Paulsen, P., Kanzler, P., Hilbert, F., Mayrhofer, S., Baumgartner, S., Smulders, F.J.M., 2005. Comparison of three methods for detecting *Campylobacter* spp. in chilled or frozen meat. International Journal of Food Microbiology 103, 229-233.

236. Perni, S., Chalise, P.R., Shama, G., Kong, M.G., 2007. Bacterial cells exposed to nanosecond pulsed electric fields show lethal and sublethal effects. International Journal of Food Microbiology 120, 311-314.

237. Perry, J.D., Freydiere, A.M., 2007. The application of chromogenic media in clinical microbiology. Journal of Applied Microbiology 103, 2046-2055.

238. Porebski, S., Bailey, L.G., Baum, B.R., 1997. Modification of a CTAB DNA extraction protocol for plants containing high polysaccharide and polyphenol components. Plant Molecular Biology Reporter 15, 8-15.

239. Posse, B., De Zutter, L., Heyndrickx, M., Herman, L., 2008. Novel differential and confirmation plating media for Shiga toxin-producing *Escherichia coli* serotypes O26, O103, O111, O145 and sorbitol-positive and -negative O157. Fems Microbiology Letters 282, 124-131.

240. Raccach, M., Bamiro, T., Clinch, J., Combs, G., Gierczynski, A., Karam, R., 2002. Frozen storage of *Escherichia coli* O157:H7 in vegetable broth. Food Control 13, 381-385.

241. Rahman, M.S., 1999. Handbook of Food Preservation. Marcel Dekkers, Inc., New York, USA, 1-889.

242. Ramesh, A., Padmapriya, B.P., Chandrashekar, A., Varadaraj, M.C., 2002. Application of a convenient DNA extraction method and multiplex PCR for the direct detection of *Staphylococcus aureus* and *Yersinia enterocolitica* in milk samples. Molecular and Cellular Probes 16, 307-314.

243. Rangel, J.M., Sparling, P.H., Crowe, C., Griffin, P.M., Swerdlow, D.L., 2005. Epidemiology of *Escherichia coli* O157:H7 outbreaks, United States, 1982-2002. Emerging Infectious Diseases 11, 603-609.

244. Raso, J., Barbosa-Canovas, G.V., 2003. Nonthermal preservation of foods using combined processing techniques. Critical Reviews in Food Science and Nutrition 43, 265-285.

245. Rasooly, A., Herold, K.E., 2006. Biosensors for the analysis of food- and waterborne pathogens and their toxins. Journal of Aoac International 89, 873-883.

246. Ray, B., Speck, M.L., 1972. Repair of injury induced by freezing *Escherichia coli* as influenced by recovery medium. Applied Microbiology 24, 258-263.

247. Rees, C.E.D., Dodd, C.E.R., 2006. Phage for rapid detection and control of bacterial pathogens in food. Advances in Applied Microbiology 59, 159-186.

248. Rentenaar, I.M.F., 1996. MicroVal, a challenging Eureka project. Food Control 7, 31-36.

249. Restaino, L., Frampton, E.W., Spitz, H., 2001. Repair and growth of heat- and freeze-injured *Escherichia coli* O157:H7 in selective enrichment broths. Food Microbiology 18, 617-629.

250. Ribeiro, M.H., Manha, S., Brito, L., 2006. The effects of salt and pH stress on the growth rates of persistent strains of *Listeria monocytogenes* collected from specific ecological niches. Food Research International 39, 816-822.

251. Ricci, F., Volpe, G., Micheli, L., Palleschi, G., 2007. A review on novel developments and applications of immunosensors in food analysis. Analytica chimica acta 111-129.

252. Rijpens, N., Herman, L., 2004. Comparison of selective and nonselective primary enrichments for the detection of *Listeria monocytogenes* in cheese. International Journal of Food Microbiology 94, 15-22.

253. Rijpens, N., Jannes, G., Herman, L., 2002. Messenger RNA-based RT-PCR detection of viable *Salmonella*. International Dairy Journal 12, 233-238.

254. Roberts T.A., Baird-Parker, A.C., Tompkin, R.B. Micro-organisms in Foods 5 - Microbiological specifications of food pathogens - ICMSF (1st edition).Blackie Academic & Professional, Londen, 1-513.

255. Rocourt, J., Jacquet, C., Reilly, A., 2000. Epidemiology of human listeriosis and seafoods. International Journal of Food Microbiology 62, 197-209.

256. Rodriguez-Lazaro, D., Hernandez, M., 2006. Isolation of *Listeria monocytogenes* DNA from meat products for quantitative detection by real-time PCR. Journal of Rapid Methods and Automation in Microbiology 14, 395-404.

257. Rodriguez-Lazaro, D., Lombard, B., Smith, H., Rzezutka, A., D'Agostino, M., Helmuth, R., Schroeter, A., Malorny, B., Miko, A., Guerra, B., Davison, J., Kobilinsky, A., Hernandez, M., Bertheau, Y., Cook, N., 2007. Trends in analytical methodology in food safety and quality: monitoring microorganisms and genetically modified organisms. Trends in Food Science & Technology 18, 306-319.

258. Rossmanith, P., Suss, B., Wagner, M., Hein, I., 2007. Development of matrix lysis for concentration of gram positive bacteria from food and blood. Journal of Microbiological Methods 69, 504-511.

259. Rotariu, O., Ogden, I.D., MacRae, M., Badescu, V., Strachan, N.J.C., 2005. An immunomagnetic separator for concentration of pathogenic micro-organisms from large volume samples. Journal of Magnetism and Magnetic Materials 293, 589-596.

260. Ruano-Lopez, J.M., Agirregabiria, M., Olabarria, G., Verdoy, D., Bang, D.D., Bu, M.Q., Wolff, A., Voigt, A., Dziuban, J.A., Walczak, R., Berganzo, J., 2009. The SmartBioPhone (TM), a point of care vision under development through two European projects: OPTOLABCARD and LABONFOIL. Lab on A Chip 9, 1495-1499.

261. Ryser E.T. and Marth E.H., 1999. *Listeria*, Listeriosis and Food Safety (2nd edition). Marcel Dekker, Inc., New York, 1-738.

262. Safarik, I., Safarikova, M., Forsythe, S.J., 1995. The application of magnetic separations in applied microbiology. Journal of Applied Bacteriology 78, 575-585.

263. Samkutty, P.J., Gough, R.H., Adkinson, R.W., McGrew, P., 2001. Rapid assessment of the bacteriological quality of raw milk using ATP bioluminescence. Journal of Food Protection 64, 208-212.

264. Sampers, I., Habib, I., Berkvens, D., Dumoulin, A., De Zutter, L., Uyttendaele, M., 2008. Processing practices contributing to *Campylobacter* contamination in Belgian chicken meat preparations. International Journal of Food Microbiology 128, 297-303.

265. Scheu, P., Gasch, A., Berghof, K., 1999. Rapid detection of *Listeria monocytogenes* by PCR-ELISA. Letters in Applied Microbiology 29, 416-420.

266. Schmid, M.W., Lehner, A., Stephan, R., Schleifer, K.H., Meier, H., 2005. Development and application of oligonucleotide probes for *in situ* detection of thermotolerant *Campylobacter* in chicken faecal and liver samples. International Journal of Food Microbiology 105, 245-255.

267. Schonenbrucher, V., Mallinson, E.T., Bulte, M., 2008. A comparison of standard cultural methods for the detection of foodborne *Salmonella* species including three new chromogenic plating media. International Journal of Food Microbiology 123, 61-66.

268. Schulten, S.M., In't Veld, P.H., Nagelkerke, N.J.D., Scotter, S., de Buyser, M.L., Rollier, P., Lahellec, C., 2000. Evaluation of the ISO 7932 standard for the enumeration of *Bacillus cereus* in foods. International Journal of Food Microbiology 57, 53-61.

269. Seo, K.H., Brackett, R.E., Frank, J.F., Hilliard, S., 1998. Immunomagnetic separation and flow cytometry for rapid detection of *Escherichia coli* O157:H7. Journal of Food Protection 61, 812-816.

270. Settanni, L., Corsetti, A., 2007. The use of multiplex PCR to detect and differentiate food- and beverage-associated microorganisms: A review. Journal of Microbiological Methods 69, 1-22.

271. Sewell, A.M., Warburton, D.W., Boville, A., Daley, E.F., Mullen, K., 2003. The development of an efficient and rapid enzyme linked fluorescent assay method for the detection of *Listeria* spp. from foods. International Journal of Food Microbiology 81, 123-129.

272. Shabala, L., Budde, B., Ross, T., Siegumfeldt, H., McMeekin, T., 2002. Responses of *Listeria monocytogenes* to acid stress and glucose availability monitored by measurements of intracellular pH and viable counts. International Journal of Food Microbiology 75, 89-97.

273. Siegumfeldt, H., Rechinger, K.B., Jakobsen, M., 1999. Use of fluorescence ratio imaging for intracellular pH determination of individual bacterial cells in mixed cultures. Microbiology-UK 145, 1703-1709.

274. Siegumfeldt, H., Rechinger, K.B., Jakobsen, M., 2000. Dynamic changes of intracellular pH in individual lactic acid bacterium cells in response to a rapid drop in extracellular pH. Applied and Environmental Microbiology 66, 2330-2335.

275. Siragusa, G.R., Cutter, C.N., Dorsa, W.J., Koohmaraie, M., 1995. Use of a rapid microbial ATP bioluminescence assay to detect contamination on beef and pork carcasses. Journal of Food Protection 58, 770-775.

276. Skottrup, P.D., Nicolaisen, M., Justesen, A.F., 2008. Towards on-site pathogen detection using antibody-based sensors. Biosensors & Bioelectronics 24, 339-348.

277. Smigic, N., Rajkovic, A., Antal, E., Medic, H., Lipnicka, B., Uyttendaele, M., Devlieghere, F., 2009a. Treatment of *Escherichia coli* O157:H7 with lactic acid, neutralized electrolyzed oxidizing water and chlorine dioxide followed by growth under sub-optimal conditions of temperature, pH and modified atmosphere. Food Microbiology 26, 629-637.

278. Smigic, N., Rajkovic, A., Nielsen, D.S., Siegumfeldt, H., Uyttendaele, M., Devlieghere, F., Arneborg, N., 2009b. Intracellular pH as an indicator of viability and resuscitation of *Campylobacter jejuni* after decontamination with lactic acid. International Journal of Food Microbiology 135, 136-143.

279. Smith, J.L., 1995. Arthritis, Guillain-Barre-Syndrome, and other sequelae of *Campylobacter jejuni* enteritis. Journal of Food Protection 58, 1153-1170.

280. Starbuck, M.A.B., Hill, P.J., Stewart, G.S.A.B., 1992. Ultra sensitive detection of *Listeria monocytogenes* in milk by the Polymerase Chain-Reaction (PCR). Letters in Applied Microbiology 15, 248-252.

281. Stephan, R., Schumacher, S., Zychowska, A., 2002. System for detection of *Listeria monocytogenes* in minced meat and environment samples of food processing plants - Evaluation of the VIT technology as a rapid method. Fleischwirtschaft 83, 94-96.

282. Stephan, R., Schumacher, S., Zychowska, M.A., 2003. The VIT (R) technology for rapid detection of *Listeria monocytogenes* and other *Listeria* spp. International Journal of Food Microbiology 89, 287-290.

283. Stephens, P.J., Joynson, J.A., 1998. Direct inoculation into media containing bile salts and antibiotics is unsuitable for the detection of acid/salt stressed *Escherichia coli* O157:H7. Letters in Applied Microbiology 27, 147-151.

284. Stephens, P.J., Joynson, J.A., Davies, K.W., Holbrook, R., LappinScott, H.M., Humphrey, T.J., 1997. The use of an automated growth analyser to measure recovery times of single heat-injured *Salmonella* cells. Journal of Applied Microbiology 83, 445-455.

285. Stessl, B., Luf, W., Wagner, M., Schoder, D., 2009. Performance testing of six chromogenic ALOA-type media for the detection of *Listeria monocytogenes*. Journal of Applied Microbiology 106, 651-659.

286. Stevens, K.A., Jaykus, L.A., 2004. Bacterial separation and concentration from complex sample matrices: A review. Critical Reviews in Microbiology 30, 7-24.

287. Stewart, G.S.A.B., Jassim, S.A.A., Denyer, S.P., Newby, P., Linley, K., Dhir, V.K., 1998. The specific and sensitive detection of bacterial

pathogens within 4 h using bacteriophage amplification. Journal of Applied Microbiology 84, 777-783.

288. Stewart, G.S.A.B., Williams, P., 1992. *Lux* genes and the applications of bacterial bioluminescence. Journal of General Microbiology 138, 1289-1300.

289. Suhren, G., Reichmuth, J., 1997. Determination of the bacteriological quality of raw milk by automatic fluorescent microscopic counting of single bacteria (Bactoscan 8000) - a review. Kieler Milchwirtschaftliche Forschungsberichte 49, 163-186.

290. Suhren, G., Walte, H.G., 1998. First experiences with automatic flow cytometric determination of total bacterial count in raw milk. Kieler Milchwirtschaftliche Forschungsberichte 50, 249-275.

291. Suzuki, S., Shibata, N., Yamane, K., Wachino, J., Ito, K., Arakawa, Y., 2009. Change in the prevalence of extended-spectrum-beta-lactamase-producing *Escherichia coli* in Japan by clonal spread. Journal of Antimicrobial Chemotherapy 63, 72-79.

292. Szabo, I., Scherer, K., Roesler, U., Appel, B., Nockler, K., Hensel, A., 2008. Comparative examination and validation of ELISA test systems for *Salmonella typhimurium* diagnosis of slaughtering pigs. International Journal of Food Microbiology 124, 65-69.

293. Tanaka, Y., Yamaguchi, N., Nasu, M., 2000. Viability of *Escherichia coli* O157:H7 in natural river water determined by the use of flow cytometry. Journal of Applied Microbiology 88, 228-236.

294. Teunis, P., Takumi, K., Shinagawa, K., 2004. Dose response for infection by *Escherichia coli* O157:H7 from outbreak data. Risk Analysis 24, 401-407.

295. Thacker, J.D., Casale, E.S., Tucker, C.M., 1996. Immunoassays (ELISA) for rapid, quantitative analysis in the food-processing industry. Journal of Agricultural and Food Chemistry 44, 2680-2685.

296. Thevenot, D., Dernburg, A., Vernozy-Rozand, C., 2006. An updated review of *Listeria monocytogenes* in the pork meat industry and its products. Journal of Applied Microbiology 101, 7-17.

297. Tilden, J., Young, W., McNamara, A.M., Custer, C., Boesel, B., LambertFair, M., Majkowski, J., Vugia, D., Werner, S.B., Hollingsworth, J., Morris, J.G., 1996. A new route of transmission for *Escherichia coli*: Infection from dry fermented salami. American Journal of Public Health 86, 1142-1145.

298. Tokarskyy, O., Marshall, D.L., 2008. Immunosensors for rapid detection of *Escherichia coli* O157:H7 - Perspectives for use in the meat processing industry. Food Microbiology 25, 1-12.

299. Torresvitela, M.A.R., Escartin, E.F., Castillo, A., 1995. Risk of salmonellosis associated with consumption of chocolate in Mexico. Journal of Food Protection 58, 478-481.

300. Tortajada, M., Martinez-Culebras, P.V., Navarro, V., Monzo, H., Ramon, D., 2009. Evaluation of DNA extraction methods for PCR detection of fungal and bacterial contamination in cocoa extracts. European Food Research and Technology 230, 79-87.

301. Urfer, E., Rossier, P., Mean, F., Krending, M.J., Burnens, A., Bille, J., Francioli, P., Zwahlen, A., 2000. Outbreak of *Salmonella braenderup* gastroenteritis due to contaminated meat pies: clinical and molecular epidemiology. Clinical Microbiology and Infection 6, 536-542.

302. Uyttendaele, M., Baert, K., Ghafir, Y., Daube, G., De Zutter, L., Herman, L., Dierick, K., Pierard, D., Dubois, J.J., Horion, B., Debevere, J., 2006.

Quantitative risk assessment of *Campylobacter* spp. in poultry based meat preparations as one of the factors to support the development of risk-based microbiological criteria in Belgium. International Journal of Food Microbiology 111, 149-163.

303. Uyttendaele, M., Bagamboula, C.F., De Smet, E., Van Wilder, S., Debevere, J., 2001a. Evaluation of culture media for enrichment and isolation of *Shigella sonnei* and *S. flexneri*. International Journal of Food Microbiology 70, 255-265.

304. Uyttendaele, M., Debevere, J., Lindqvist, R., 1999. Evaluation of buoyant density centrifugation as a sample preparation method for NASBA-ELGA detection of *Campylobacter jejuni* in foods. Food Microbiology 16, 575-582.

305. Uyttendaele, M., Taverniers, I., Debevere, J., 2001b. Effect of stress induced by suboptimal growth factors on survival of *Escherichia coli* O157:H7. International Journal of Food Microbiology 66, 31-37.

306. Uyttendaele, M., Van Hoorde, I., Debevere, J., 2000. The use of immuno-magnetic separation (IMS) as a tool in a sample preparation method for direct detection of *L. monocytogenes* in cheese. International Journal of Food Microbiology 54, 205-212.

307. Uyttendaele, M., Vankeirsbilck, S., Debevere, J., 2001c. Recovery of heat-stressed *E. coli* O157:H7 from ground beef and survival of *E. coli* O157:H7 in refrigerated and frozen ground beef and in fermented sausage kept at 7 degrees C and 22 degrees C. Food Microbiology 18, 511-519.

308. Uyttendaele, M., Vanwildemeersch, K., Debevere, J., 2003. Evaluation of real-time PCR vs automated ELISA and a conventional culture method using a semi-solid medium for detection of *Salmonella*. Letters in Applied Microbiology 37, 386-391.

309. Van Griethuysen, A., Bes, M., Etienne, J., Zbinden, R., Kluytmans, J., 2001. International multicenter evaluation of latex agglutination tests for identification of *Staphylococcus aureus*. Journal of Clinical Microbiology 39, 86-89.

310. Van Houteghem, N., Devlieghere, F., Rajkovic, A., Gomez, S.M.O., Uyttendaele, M., Debevere, J., 2008. Effects of CO2 on the resuscitation of *Listeria monocytogenes* injured by various bactericidal treatments. International Journal of Food Microbiology 123, 67-73.

311. Vellinga, A., Van Loock, F., 2002. The dioxin crisis as experiment to determine poultry-related *Campylobacter* enteritis. Emerging Infectious Diseases 8, 19-22.

312. Vermeiren, L., Devlieghere, F., Vandekinderen, I., Rajtak, U., Debevere, J., 2006. The sensory acceptability of cooked meat products treated with a protective culture depends on glucose content and buffering capacity: A case study with *Lactobacillus sakei* 10A. Meat Science 74, 532-545.

313. Vernozy-Rozand, C., Feng, P., Montet, M.P., Ray-Gueniot, S., Villard, L., Bavai, C., Meyrand, A., Mazuy, C., Atrache, V., 2000. Detection of *Escherichia coli* O157:H7 in heifers' faecal samples using an automated immunoconcentration system. Letters in Applied Microbiology 30, 217-222.

314. Vieira-Pinto, M., Oliveira, M., Bernardo, F., Martins, C., 2007. Rapid detection of *Salmonella* sp in pork samples using fluorescent in situ hybridization: a comparison with VIDAS (R)-SLM system and ISO 6579 cultural method. Arquivo Brasileiro de Medicina Veterinaria e Zootecnia 59, 1388-1393.

315. Vlaemynck, G.M., Moermans, R., 1996. Comparison of EB and Fraser enrichment broths for the detection of *Listeria* spp and *Listeria monocytogenes* in raw-milk dairy products and environmental samples. Journal of Food Protection 59, 1172-1175.

316. Wagner, M., Horn, M., Daims, H., 2003. Fluorescence *in situ* hybridisation for the identification and characterisation of prokaryotes. Current Opinion in Microbiology 6, 302-309.

317. Waldroup, A.L., 1996. Contamination of raw poultry with pathogens. Worlds Poultry Science Journal 52, 7-25.

318. Walls, I., 2005. Achieving continuous improvement in reductions in foodborne listeriosis - A risk-based approach. Journal of Food Protection 68, 1932-1994.

319. Wang, A.H., Yang, Y.H., Lu, Q., Wang, Y., Chen, Y., Deng, L., Ding, H., Deng, Q.L., Zhang, H., Wang, C.Q., Liu, L., Xu, X.W., Wang, L., Shen, X.Z., 2008. Presence of *qnr* gene in *Escherichia coli* and *Klebsiella pneumoniae* resistant to ciprofloxacin isolated from pediatric patients in China. BMC Infectious Diseases 8, 1-6.

320. Warren, B.R., Yuk, H.G., Schneider, K.R., 2007. Detection of *Salmonella* by flow-through immunocapture real-time PCR in selected foods within 8 hours. Journal of Food Protection 70, 1002-1006.

321. Wattiau, P., Weijers, T., Andreoli, P., Schliker, C., Veken, H.V., Maas, H.M.E., Verbruggen, A.J., Heck, M.E.O.C., Wannet, W.J., Imberechts, H., Vos, P., 2008. Evaluation of the Premi (R) Test *Salmonella*, a commercial low-density DNA microarray system intended for routine identification and typing of *Salmonella* enterica. International Journal of Food Microbiology 123, 293-298.

322. Weagant, S.D., Bryant, J.L., Jinneman, K.G., 1995. An improved rapid technique for isolation of *Escherichia coli* O157:H7 from foods. Journal of Food Protection 58, 7-12.

323. Welinder-Olsson, C., Kaijser, B., 2005. Enterohemorrhagic *Escherichia coli* (EHEC). Scandinavian Journal of Infectious Diseases 37, 405-416.

324. Wells, J.M., Bennik, M.H.J., 2003. Genomics of food-borne bacterial pathogens. Nutrition Research Reviews 16, 21-35.

325. Werber, D., Dreesman, J., Feil, F., van Treeck, U., Fell, G., Ethelberg, S., Hauri, A.M., Roggentin, P., Prager, R., ST Fisher, I., Behnke, S.C., Bartelt, E., Weise, E., Ellis, A., Siitonen, A., Andersson, Y., Tschape, H., Kramer, M.H., Ammon, A., 2005. International outbreak of *Salmonella Oranienburg* due to German chocolate. BMC Infectious Diseases 5, 1-10.

326. Wesche, A.M., Gurtler, J.B., Marks, B.P., Ryser, E.T., 2009. Stress, sublethal injury, resuscitation, and virulence of bacterial foodborne pathogens. Journal of Food Protection 72, 1121-1138.

327. Wilson, I.G., 1997. Inhibition and facilitation of nucleic acid amplification. Applied and Environmental Microbiology 63, 3741-3751.

328. Wright, D.J., Chapman, P.A., Siddons, C.A., 1994. Immunomagnetic Separation as a sensitive method for isolating *Escherichia coli* O157 from food samples. Epidemiology and Infection 113, 31-39.

329. Wu, V.C.H., 2008. A review of microbial injury and recovery methods in food. Food Microbiology 25, 735-744.

330. Wu, V.C.H., Gill, V., Oberst, R., Phebus, R., Fung, D.Y.C., 2004. Rapid protocol (5.25 H) for the detection of *Escherichia coli* O157:H7 in raw ground beef by an immuno-capture system (Pathatrix) in combination with Colortrix and CT-SMAC. Journal of Rapid Methods and Automation in Microbiology 12, 57-67.

331. Wuytack, E.Y., Phuong, L.D.T., Aertsen, A., Reyns, K.M.F., Marquenie, D., De Ketelaere, B., Masschalck, B., Van Opstal, B.M.I., Diels, A.M.J.,

Michiels, C.W., 2003. Comparison of sublethal injury induced in *Salmonella enterica* serovar *Typhimurium* by heat and by different nonthermal treatments. Journal of Food Protection 66, 31-37.

332. Xicohtencati-Cortes, J., Monteiro-Neto, V., Ledesma, M.A., Jordan, D.M., Francetic, O., Kaper, J.B., Puente, J.L., Giron, J.A., 2007. Intestinal adherence associated with type IV pili of enterohemorrhagic *Escherichia coli* O157:H7. Journal of Clinical Investigation 117, 3519-3529.

333. Yamasaki, M., Igimi, S., Katayama, Y., Yamamoto, S., Amano, F., 2004. Identification of an oxidative stress-sensitive protein from *Campylobacter jejuni*, homologous to rubredoxin oxidoreductase/rubrerythrin. Fems Microbiology Letters 235, 57-63.

334. Yang, H., Li, H.P., Jiang, X.P., 2008. Detection of foodborne pathogens using bioconjugated nanomaterials. Microfluidics and Nanofluidics 5, 571-583.

335. Yeh, K.S., Tsai, C.E., Chen, S.P., Liao, C.W., 2002. Comparison between VIDAS automatic enzyme-linked fluorescent immunoassay and culture method for *Salmonella* recovery from pork carcass sponge samples. Journal of Food Protection 65, 1656-1659.

336. Yoon, J.W., Hovde, C.J., 2008. All blood, no stool: enterohemorrhagic *Escherichia coli* O157:H7 infection. Journal of Veterinary Science 9, 219-231.

337. Zadik, P.M., Chapman, P.A., Siddons, C.A., 1993. Use of tellurite for the selection of verocytotoxigenic *Escherichia coli* O157. Journal of Medical Microbiology 39, 155-158.

338. Zapparoli, G., 2004. Colony dimorphism associated with stress resistance in *Oenococcus oeni* VP01 cells during stationary growth phase. Fems Microbiology Letters 239, 261-265.

339. Zhang, X.Q., Guo, Q., Cui, D.X., 2009. Recent advances in nanotechnology applied to biosensors. Sensors 9, 1033-1053.

340. Zhou, J.H., 2003. Microarrays for bacterial detection and microbial community analysis. Current Opinion in Microbiology 6, 288-294.

341. Zhu, M.J., Du, M., Cordray, J., Ahn, D.U., 2005. Control of *Listeria monocytogenes* contamination in ready-to-eat meat products. Comprehensive Reviews in Food Science and Food Safety 4, 34-42.

Curriculum vitae

Curriculum vitae

Vicky Jasson werd geboren op 16 februari 1981 te Wilrijk. In 1999 behaalde zij het diploma Wetenschappen-Wiskunde aan het Stedelijk Lyceum Jan de Voslei te Antwerpen. In 2005 promoveerde zij tot Bio-ingenieur in de Cel- en Genbiotechnologie aan de Universiteit Gent.

In september 2005 startte zij als doctoraatsbursaal op het laboratorium voor Levensmiddelenmicrobiologie en –conservering aan de Faculteit Bio-ingenieurswetenschappen, Universiteit Gent. Haar onderzoeksproject kaderde in werkpakket 6 van het Europees project Pathogen Combat (project no:FOOD-CT-2005-07081). In kader van dit onderzoek werd de resuscitatie van subletaal beschadigde cellen van verschillende voedselgebonden pathogenen onderzocht. Zij werkte aan dit doctoraat onder leiding van haar promotoren Prof. dr. ir. Mieke Uyttendaele en Em. Prof. dr. ir. Johan Debevere, beide verbonden aan de UGent.

Tijdens haar onderzoek publiceerde zij in internationale tijdschriften, en nam zij deel aan internationale en nationale workshops, symposia en congressen. Verder begeleidde zij binnen- en buitenlandse thesisstudenten en werkte zij mee aan de practica voor de vakken ‘Levensmiddelenmicrobiologie en -conservering’, ‘Moleculair-microbiële technieken’ en ‘Microbiologische veiligheid en analyse van levensmiddelen’ in de opleiding Bio-ingenieur en het vak ‘Food microbiology’ in de Master opleiding Food Science and Nutrition. Eveneens verzorgde zij lessen voor het ‘International Training Program in Food Safety, Quality Assurance and Risk Analysis’ georganiseerd door de laboratoria voor Levensmiddelenmicrobiologie en –conservering en Levensmiddelenchemie en humane voeding. Ze was ook betrokken bij dienstverlening waarbij zij onder andere training heeft gegeven over ‘microbiologische analyse van levensmiddelen’ voor industriële partners. Ook werd zij gevraagd peer-reviewer te zijn voor International Journal of Food Microbiology.

Curriculum vitae

Vicky Jasson was born in Wilrijk on February the 16th in 1981. In 1999 she obtained a degree in Science and Mathematics at Stedelijk Lyceum Jan de Voslei in Antwerp. In 2005 she graduated as Bioscience Engineer in the Cell and Gene Biotechnology at the University of Ghent.

In September 2005 she started her PhD study at the Laboratory of Food Safety and Food Preservation at the Faculty of Bioscience Engineering, University of Ghent. Her PhD research was situated in work package 6 of the European project Pathogen Combat (project no:FOOD-CT-2005-07081). In the framework of this research focus was set on resuscitation and enrichment of food-borne pathogens. Vicky Jasson worked on this topic under the guidance of two promoters Prof. dr. ir. Mieke Uyttendaele and Emeritus Prof. dr. ir. Johan Debevere both associated with the University of Ghent.

During her PhD study she published in several international journals, attended different national and international workshops, symposia and congresses. She also guided Belgian and foreign thesis students and was involved with practical sessions of the courses: 'Levensmiddelenmicrobiologie en -conservering', 'Moleculair-microbiële technieken' and 'Microbiologische veiligheid en analyse van levensmiddelen' for the Master Bioscience Engineering and the course 'Food microbiology' in the Master Food Science and Nutrition. She also gave some courses of the 'International Training Program in Food Safety, Quality Assurance and Risk Analysis' organized by the Laboratory of Food Microbiology and Food Preservation and the and laboratory of Food chemistry and human nutrition. She contributed to services such as providing training with regard to 'microbial analysis of food' for industrial partners. International Journal for Food Microbiology asked her to be a peer-reviewer.

Publication list

Publications in A1 peer-reviewed journals

- Jasson, V., Uyttendaele, M., Rajkovic, A. and Debevere, J. (2007). Establishment of procedures provoking sub-lethal injury of *Listeria monocytogenes*, *Campylobacter jejuni* and *Escherichia coli* O157 to serve method performance testing. International Journal of Food Microbiology, 18, 241-249.
- Jasson, V., Rajkovic, A., Debevere, J., Uyttendaele, M. (2009). Kinetics of resuscitation and growth of *L. monocytogenes* as a tool to select appropriate enrichment conditions as a prior step to rapid detection methods. Food Microbiology, 26, 88-93.
- Jasson, V., Rajkovic, A., Baert, L., Debevere, J., Uyttendaele, M. (2009). Comparison of enrichment conditions for rapid detection of low numbers of sub-lethally injured *Escherichia coli* O157 in food. Journal of Food Protection, 72, 1862-1868.
- Jasson, V., Sampers, I., Botteldoorn, N., López-Gálvez, F., Baert, L., Denayer, S., Rajkovic, A., Habib, I., De Zutter, L., Debevere, J., Uyttendaele, M. (2009). Characterization of *Escherichia coli* from raw poultry in Belgium and impact on the detection of *Campylobacter jejuni* using Bolton broth. International Journal of Food Microbiology, 135, 248-253.
- Jasson, V. Jacxsens, L., Luning, P., Rajkovic, A. Uyttendaele M. (2009). Alternative microbial methods: An overview and selection criteria. Food Microbiology, submitted.
- Jasson, V. Baert, L., Uyttendaele M. (2009). Appropriate sample preparation for reliable real-time PCR detection of low numbers of healthy and sub-lethally injured *Salmonella enterica* in chocolate. Letters in Applied Microbiology, submitted.
- Uyttendale M., Rajkovic, A., Habib, I., Baert, L., Jasson, V., Rantsiou, K., Cocolin, L. (2009). Use of molecular techniques in food safety diagnostics. International Journal of Food Microbiology, submitted.

Extended abstracts of symposia and workshops

- Jasson, V., Uyttendaele, M. and Rajkovic, A. (2007). Optimization of resuscitation and capture as a prerequisite for fast detection of *L. monocytogenes* in food samples. Poster presentation, Twelfth Conference on Food Microbiology, 21-22 juni 2007, Liege, Belgium, 103.
- Rajkovic, A., Smigic, N., Jasson, V., Uyttendaele, M., and Devlieghere, F. (2008). Sublethal food processing: effect on the behaviour of pathogens under suboptimal conditions. Invited lecture. Fourth Croatian Congress of Microbiology with International Participation, September 24th-27th, 2008, Zadar, Croatia.
- Jasson, V., Uyttendaele, M., Rajkovic, A., Baert, L. and Debevere, J. (2008). Appropriate enrichment conditions, a prerequisite for rapid and reliable detection of sub-lethally injured *E. coli* O157 in food using capture and real-time PCR. Oral presentation. FoodMicro 2008, September 1st – September 4th, 2008, Aberdeen, Scotland, 99.
- Jasson, V., Rajkovic, A., Miller, S., Uyttendaele, M. (2008). Appropriate enrichment conditions as a prior step for fast detection of *Listeria monocytogenes*. Poster Presentation. Workshop: New Trends in Emerging Pathogens. March 10th, 2008, Hohenhein, Germany.
- Jasson, V., Rajkovic, A., Miller, S. and Uyttendaele, M. (2008). Rapid detection of low numbers of *Listeria monocytogenes*: capture based on phage proteins and magnetic separation. Poster presentation, Rapid Methods EUROPE 2008, January 21 – January 23, 2008, Noordwijkerhout, The Netherlands.
- Jasson, V., Sampers, I., Botteldoorn, N., López-Gálvez, F., Baert, L., Denayer, S., Rajkovic, A., Habib, I., De Zutter, L., Debevere, J., Uyttendaele, M. (2009). ESBL producing *E. coli* interfering in the detection of *Campylobacter* spp. from chicken meat preparations in Belgium, Fourteenth Conference on Food Microbiology, 18-19 June 2009, Liege, Belgium, 104.
- Jasson, V., Rajkovic, A. and Uyttendaele, M. (2009). Comparison of three pillars for detection of *Listeria monocytogenes*, *Escherichia coli* O157 and *Campylobacter jejuni* within 24 hours time frame. Poster presentation, Rapid

Methods EUROPE 2009, January 26 – January 28, 2009, Noordwijkerhout, The Netherlands.

- Jasson, V. Jacxsens, L., Rajkovic,, A., Uyttendaele, M. A selection tool for application of the most appropriate microbial method of analysis. Poster presentation. IAFP European Symposium 2009, October 7th - October 9th, 2009, Berlin, Germany.